# CONTENTS

*Preface to the Second Edition*       *page* vii

*Preface to the First Edition*       ix

*Author's Note*       x

1    FIELD OBSERVATIONS OF MIGRATION       1

Examples of migration. Ringing. Visible and audible migration. Radar. Form and process of migration. Probable use of astronomical clues.

2    EXPERIMENTAL EVIDENCE FOR BEARING-AND-DISTANCE NAVIGATION       10

Transplanting of young birds. Adoption of foster homes. Displacement of migrants. Inborn nature of migration orientation. Directional training in Pigeons. 'Nonsense' orientation.

3    THE PHYSICAL BASES OF DIURNAL ONE-DIRECTION NAVIGATION       22

Use of caged migrants. Automatic registration. Possible effects of Earth's magnetic field and rotation. Sun-compass orientation. Artificial sun. Time compensation. Clock shifting. Shifts in longitude and latitude.

4    THE PHYSICAL BASES OF NOCTURNAL ONE-DIRECTION NAVIGATION       41

Utilization of stars. Independence from time. Orientation by star pattern. Interrelations between time, longitude and season. Planetarium experiments. Evolution. Moon orientation.

5    HOMING EXPERIMENTS       54

Pigeons. Variation in ability. Seasonal effects. Non-breeding wild birds. Breeding wild birds. Outstanding feats. Conditions for fast homing. Individual variations. Random and biased search. Determination of homeward track. Aerial pursuit.

6    HOMING ORIENTATION       76

Techniques. Pigeons. Time required for orientation. Confusion with 'nonsense' orientation. Effect of distance. Sun-compass and landmarks. Inborn ability. Limitation of experience. Wild birds, free-flying and caged.

7    THEORIES OF SENSORY CONTACT WITH HOME
     AND OF INERTIAL NAVIGATION      *page* 94

Direct vision. Atmospheric sensitivity. Extrasensory perception. Inertial navigation theory. Practical tests. Stable 'instrument bed'.

8    THEORIES OF NAVIGATION BY
     GEOPHYSICAL 'GRIDS'      101

Grid navigation. Earth rotation. Coriolis force. Earth's magnetic field. Tests of magnetic sensitivity. Conjugate point experiments. Radio and radar transmissions. Unknown factors.

9    THEORIES OF NAVIGATION BY
     ASTRONOMICAL 'GRIDS'      112

Development of theories. Evidence for use of sun. Sun co-ordinates versus sun-compass. Planetarium evidence for use of star co-ordinates. Basic theory of astronavigation.

10    THEORIES OF NAVIGATION BY A 'GRID'
     DERIVED FROM THE SUN'S CO-ORDINATES      122

Sun arc extrapolation hypothesis. Theoretical difficulties. Rate-of-change of altitude hypothesis. Ambiguities. The two hypotheses compared.

11    FIELD TESTS OF THEORIES OF NAVIGATION
     BY THE SUN'S CO-ORDINATES      132

Making use of seasonal changes in sun altitude. Conflicting results. Impairment of homing by restricting view of horizon. Effect of massive time shifts. Conflicting results. Effect of small time shifts.

12    THE ANATOMICAL AND PHYSIOLOGICAL
     LIMITATIONS OF THE AVIAN EYE      142

Movement detection and measurement. Extrapolation. Retinal structure. Dioptric apparatus. Resolution. Dark adaptation.

13    MOTION, TIME AND MEMORY      152

Stability of head. Blind flying. Gravity receptors. Temporal stereotypy of movements and calls. Circadian rhythms. Accuracy versus rigidity. Memory excellence. Functions of bi-coordinate navigation.

*Scientific names of species mentioned*      164

*References*      167

*Index*      193

CAMBRIDGE MONOGRAPHS IN
EXPERIMENTAL BIOLOGY
No. 3

EDITORS:
P. W. BRIAN, G. M. HUGHES
GEORGE SALT (*General Editor*)
E. N. WILLMER

# BIRD NAVIGATION

# THE SERIES

1   V. B. WIGGLESWORTH. The Physiology of Insect Metamorphosis

2   G. H. BEALE. The Genetics of *Paramecium aurelia*

3   G. V. T. MATTHEWS. Bird Navigation. 2nd edition

4   A. D. LEES. The Physiology of Diapause in Arthropods

5   E. B. EDNEY. The Water-relations of Terrestrial Arthropods

6   LILLIAN E. HAWKER. The Physiology of Reproduction in Fungi

7   R. A. BEATTY. Parthenogenesis and Polyploidy in Mammalian Development

8   G. HOYLE. Comparative Physiology of the Nervous Control of Muscular Contraction

9   J. W. S. PRINGLE. Insect Flight

10  D. B. CARLISLE and SIR FRANCIS KNOWLES. Endocrine Control in Crustaceans

11  DAVID D. DAVIES. Intermediary Metabolism in Plants

12  W. H. THORPE. Bird-song

13  JANET E. HARKER. The Physiology of Diurnal Rhythms

14  J. E. TREHERNE. The Neurochemistry of Arthropods

15  R. N. ROBERTSON. Protons, Electrons, Phosphorylation and Active Transport

# BIRD NAVIGATION

BY

G. V. T. MATTHEWS, M.A., PH.D.

*The Wildfowl Trust, Slimbridge*

SECOND EDITION

CAMBRIDGE
AT THE UNIVERSITY PRESS
1968

Published by the Syndics of the Cambridge University Press
Bentley House, 200 Euston Road, London, N.W. 1
American Branch: 32 East 57th Street, New York, N.Y. 10022

This edition © Cambridge University Press 1968

Library of Congress Catalogue Card Number: 68–23181
Standard Book Number: 521 07271 9 clothbound
521 09541 7 paperback

First published 1955
Second edition 1968

The royalties on this book are being given to the
Wildfowl Trust, Slimbridge

Printed in Great Britain
at the University Printing House, Cambridge
(Brooke Crutchley, University Printer)

# PREFACE TO THE SECOND EDITION

The first edition of 'Bird Navigation' preceded a period of intensive activity which threw everything back in the melting pot. When the book sold out three years later the Cambridge University Press suggested that it should be reprinted. But it had already become so out of date that I asked to be allowed to revise it extensively. The Press, for whose forbearance I am very grateful, agreed to put scientific considerations before commercial ones.

A measure of the intensification of research in this field is given by the list of references. This includes some 350 post-1954 titles as against 220 earlier ones. The latter remain after heavy pruning carried out in the interests of keeping this a slim volume.

The text has been completely re-written and I have endeavoured to present a reasonably unbiased review of the field. It does, of course, remain a personal interpretation and statement.

A particularly hard blow to development in our field was occasioned by the premature death of Gustav Kramer in 1959. His contributions need no stressing, and this book is respectfully dedicated to his memory. He would have preferred, I know, that points of controversy should not be glossed over, and I have not done so.

Finally, I must thank all those friends and colleagues who kept on pressing me to get this book finished, not least among them my wife, Janet Kear. I also owe a particular debt to Don Griffin, who has ever been a stimulation and has most kindly subjected this text to ruthless criticisms. Eleanor Temple Carrington patiently and meticulously prepared the typescript.

The work was carried out while I held a post at the Wildfowl Trust financed by the Natural Environment Research Council. I am grateful for the support and encouragement of both these bodies, and expecially for that of Peter Scott.

G. V. T. M.

*November 1967*

# PREFACE TO THE FIRST EDITION

The presumed navigational powers of animals, and in particular those of birds, have attracted scientific attention for more than a century. It is during the last ten years that a fresh impetus has been given by the advancement of new theories and the development of new experimental techniques. This monograph attempts to survey the present position which, perhaps momentarily, appears to have some coherence. The path of progress in this field is littered with discarded theories and it is possible that the one at present favoured may be found inadequate. But both the existence and the physical bases of bird navigation are now firmly established, and it is more likely that future developments will lie in a better appreciation of the way in which a bird interprets and acts upon the information available to it.

I should like to acknowledge my debt to those who both encouraged and enabled me to undertake research in this field, particularly Professor Sir James Gray, F.R.S., and Dr W. H. Thorpe, F.R.S., and to those bodies which provided financial backing, the Department of Scientific and Industrial Research and the Royal Society. I am most grateful to Dr George Salt for his help and advice in the preparation of the monograph.

G. V. T. M.

*3 November 1954*

# AUTHOR'S NOTE

In view of certain semantic controversies I would state that 'navigation' is used to mean the ability to initiate and maintain directed movement independently of learned landmarks. Directed movement with reference to the latter may be considered as 'pilotage'. 'Orientation' is not, to my mind, a substitute term for any form of navigation. It should imply the taking up of a direction in relation to a stimulus or stimuli—and no more. Qualifying adjectives should refer to the stimulus concerned, but I am aware that I have sinned by perpetrating 'nonsense' orientation. Still, no-one is likely to be misled into thinking that birds are orientating by, with or from nonsense. I have, however, eschewed 'celestial navigation', which has too much flavour of Kai Lung and would be better used in accounting for the movement of angels in pre-radar days.

Bird species are referred to by their vernacular English (or American) names, but a list of their Latin names appears at the back of the book. With so much interesting work being done with other animals, reference has perforce been made to them at various places. Here the generic name is inserted in the text to bring home the point. However, no attempt has been made to review thoroughly the massive literature on the invertebrates.

# CHAPTER I

# Field observations of migration

Bird migration has now been studied from so many angles and by so many people that it is doubtful if any one author can still do the whole subject credit, as did Landsborough Thomson (1926). Recent attempts have been made by Schüz (1952), by Dorst (1962) and by Bernis (1966). The present volume concerns itself with only one aspect, the manner in which birds find their way on their journeys, their navigation. However, as this is one of the most fascinating problems in biology, no apology need be made for such specialization.

In the last twenty years in particular there has been a very considerable amount of experimental investigation of bird navigation, and the consideration of this evidence takes up the bulk of the book. But experiments should only be embarked upon against a background of knowledge of what the animal does in its natural state. And so we must first consider the evidence provided by field observation and the collection of specimens.

These together have outlined the remarkable migrations that take place, such as that of the Bristle-thighed Curlew which nests in a coastal strip of Alaska and winters in Pacific Islands 6000 miles away, with minimal sea-crossings of 2000 miles. Although the wintering area covers an arc of 45° from the breeding grounds, the migration must at least have a strongly directional trend. The return journey from the scattered islands to the restricted breeding grounds involves a greater navigational feat, but this pales in comparison with that of the Great Shearwater. These birds range over both Atlantic Oceans up to 60° N. Yet they return in their millions to breed on the Tristan da Cunha islands lying at 40° S, spread over only 30 miles of ocean and 1500 miles from the nearest land mass.

From detailed field observations and examination of plumages we know that in many species the young of the year make their way quite independently of the adult birds. There can then be no question of the latter acting as guides. The cuckoos, of course, provide extreme examples of juvenile independence, and yet the Bronze Cuckoo migrates 2500 miles over the open sea, with minimal sea crossings of 900 miles.

When races can be distinguished in the field or hand, the movements of birds from particular areas can be followed in more detail. Strong tendencies are found for a local breeding population to reassemble in quite localized wintering areas. Thus the six sub-species of the Fox Sparrow, which breed successively down the west coast of North America, are found wintering in the reverse order to the south, the more northerly races having leap-frogged their southern neighbours. Racial discrimination produces many other examples of birds migrating much greater distances than they apparently 'need'. Thus it is the Antarctic race of the Great Skua, not the New Zealand one, which winters off Japan.

But a real upsurge in knowledge came when it was possible to follow the fates of individual birds, by marking them with metal leg bands, bearing an address and a unique number. The method has been widely adopted in the last half-century and its use has reached considerable dimensions. The total number of birds ringed is not known, but it exceeds five million for Great Britain alone. Subsequent reports of birds that have been ringed range from about 20 % in species shot for sport or as vermin, to a fraction of 1 % in the smaller passerines and in pelagic species. Allowance must be made for some bias in the proportion and location of such recoveries. They will reflect to some extent the distribution of the human population (in particular of that portion of it that is literate), and hunting practices and seasons. Again in most species the bulk of the information will relate to young birds in their first year, which will provide many more recoveries owing to their greater mortality.

On the one hand, some of the most exaggerated migrations have been confirmed, such as that of the Arctic Tern. These birds have been shown to migrate from the Canadian Arctic, where they nest within 10° of the North Pole, to the Antarctic pack-ice via the west coast of Africa. The double journey is equivalent to circling the earth at the equator. On the other hand, it has been established that after long journeys, migrants which return breed year after year in the same nest site. There have been many studies of such *Ortstreue*, diverse examples

2

being those on Bank Swallows (Stoner, 1941), various duck (Sowls, 1955), Pied Flycatchers (Haartman, 1960) and Mourning Doves (Tomlinson *et al*. 1960). When allowance is made for natural mortality it is found that practically all surviving adults return to breed in the same area the following year(s). On the other hand young birds returning after their first migration settle and breed over a much wider area. For some species the much more difficult achievement of identifying the same individual repeatedly *wintering* in a small area has been reported (e.g. Tettenborn, 1943). On a more general level the re-forming of localized breeding populations in localized wintering areas has been amply confirmed. Thus Boyd (1964) showed that Barnacle Geese, breeding in Greenland and in Spitsbergen, winter in Scotland only 100 miles apart without intermingling. A third population, breeding in Siberia winters in the north of the Netherlands. Indeed the typical migration has come to have the appearance of a 'shuttle' service between two small areas. This would require as a bare minimum of navigational equipment the ability to fly an accurate bearing-and-distance course. In many cases, that of the Great Shearwater for instance, a more precise form of navigation would seem to be needed. But not necessarily so precise as would at first sight be indicated by Richdale's (1963) finding that Sooty Shearwaters nest, after several migrations of thousands of miles from New Zealand to Japan (Phillips, 1963), an average of 7·7 feet from the point where they were originally marked. A transatlantic airline navigator does not locate his suburban villa by the methods he used to guide his aircraft from New York to London.

For many years there was great controversy as to the method of migration. There were those, following Middendorf (1855), who believed that migration took place in one general direction, on a broad front. Opposing them were the followers of Palmen (1876) who insisted that the migrants passed along certain restricted routes, on a narrow front. As is so often the case in biological controversy the correct answer is a compromise. Geyr von Schweppenburg (e.g. 1922, 1963) formulated such a theory. He suggested that there was indeed a directional trend to migration, the birds flying in a 'standard direction', typical of their particular population, while over uniform terrain or the sea. But in addition there were 'leading-lines' formed by the boundaries between favourable and unfavourable terrain, between land and sea, hill and plain, forest and savannah, desert and fruitful land. When the birds encounter such a leading-line they tend to fly along it, forming a narrow and concentrated stream just as if they were passing along one of Palmen's 'routes'. But when the obstacle is passed, or the urge to fly in the standard direction becomes paramount, the stream widens out into the broad front again.

As long as field observations were confined to counting and identifying birds that had landed or were passing within the range of unaided vision, little progress could be made in the study of the way in which migration proceeded and of the factors influencing it. We now know that such observations represent but a small and variable proportion of migratory streams. Thus, when they encounter head winds birds fly lower and are so more easily seen. Again, migrants may only appear on the ground when they have been drifted there by unusual winds and/or 'precipitated' by encountering fog or heavy cloud and rain.

The initiative for a more comprehensive study of migration was provided by Kramer (1931), who observed that birds migrating overhead by day could be detected with binoculars at considerable heights. It has since been determined that the bulk of migration occurs within a mile of the surface and is therefore accessible to this technique. The new method was seized upon with enthusiasm, and gazing through upturned glasses became widespread in North-west Europe, see, for example, Svärdson (1953) and more recently in North America (Newman & Lowery, 1962). Tinbergen (1956) and Gruys-Casimir (1965) provided a summary of daylight observations in the Netherlands, especially in so far as they illuminate the problems of bird navigation, and Perdeck (1961, 1962) has made detailed studies on Chaffinch migration.

Some evidence on the numbers and species of birds passing by night has been provided by 'kills' at lighthouses and more recently airport 'ceilometers' (Howell *et al.* 1954) and television masts (Cochran & Graber, 1958). Spectacular slaughter has been recorded, such as 50,000 birds of fifty-three species in one night (Johnston & Haines, 1957) but these happenings are abnormal both as regards the birds' behaviour and the weather conditions implicated. Attempts to probe the normal course of nocturnal migration started when Scott (1881) conceived the

4

idea of using a telescope to count birds passing across the moon's disc. Lowery (1951) took up this method and organised a network of observers in Northern America, and eventually (Lowery & Newman, 1966) published results for four nights. Strömberg (1961) used the technique in Sweden. Since the apparatus is simple and portable it is useful for small expeditions working away from base facilities, such as in Greece (Bateson & Nisbet, 1961) and Spain (Walraff & Kiepenheur, 1962). Elaborate calculations are needed to ascertain the effective size of the cone of observation which is swept across the sky, but these can be reduced to tabular form (Nisbet, 1959). A more basic limitation is that observations are confined to one-third of each month, centred on the full-moon period.

Both moon-watching by night and sky-watching by day are frustrated by cloudy weather. In places like Britain this greatly reduces their utility. Mention may be made of the technique, started by Libby (1899), of counting flight calls of migrants invisible on moonless nights or above low cloud. The unaided human ear (e.g. Ball, 1952) cannot pick up normal flight calls above 1500 feet and can give but little indication of direction. Graber & Cochran (1959, 1960) have overcome these difficulties by using a parabolic reflector and an amplifier, extending the range to 10,000 feet or more. There remains the difficulty of relating the number of calls heard to the number of birds passing, for the rate of calling undoubtedly varies between species and according to weather and time of night. However the opportunity it affords for identifying species makes it a useful accessory to other methods.

Radar has undoubtedly revolutionized the field study of bird migration and made observations almost independent of weather and fully comparable by day and by night. There was a gap between the first statement of radar's potentialities for the study of bird migration (Lack & Varley, 1945) and its actual use. Suitable microwave equipment had to be developed and research workers had to be insinuated as tolerated parasites on equipment serving airports or military requirements. The snags of the radar tool are its expense and the skilled maintenance required. Moreover, its primary users have no reverence for the 'angels' that clutter up their display screens, and strive towards technical improvements that will eliminate them. Opportunities to use such equipment may thus be passing ones, but ornithologists have seized theirs in Switzerland (Sutter, 1957; Gehring, 1963), England (Harper, 1958; Tedd & Lack, 1958; Eastwood *et al.* 1960), Sweden (Mascher *et al.* 1962), Cyprus (Adams, 1962), New England (Drury & Keith, 1962), Illinois (Graber & Hassler, 1962), Finland (Bergman & Donner, 1964) and the Mediterranean (Casement, 1966).

Two main types of radar have been employed (usually operating on 10 or 23 cm bands), the ones used at airports which give coverage within a radius of 10 miles, and the high power early warning sets which sweep a vast area, detecting small birds out to 80 miles or more. The former is the tactical instrument and offers the opportunity to investigate the nature of the echoes by visual and auditory checks. One of the main limitations of radar at present is the difficulty of identifying echoes beyond the broad categories of passerines, waders or wildfowl. Much experimental investigation of 'echo' characteristics such as that of Edwards & Houghton (1959), Gehring (1967*b*) and Schaefer (1968) is needed. The bigger set can paint in majestically a complete migratory movement, enabling a strategic appreciation of the situation as a whole. Permanent photographic records can be made of the display screens at short intervals or projected as a cine-film so that the movements of the individual echoes are readily apparent. Dyer (1967) has developed a photoelectric cell technique for analysing film of radar scans. A *caveat* must be entered concerning the quantitative analysis of radar traces. So many factors, beside the actual density of the birds, are concerned in producing the density of 'angels' observed, that only very broad categories of enumeration would seem to be justified (Nisbet, 1963*b*). Nevertheless the use of radar has built up a formidable number of data on the phenology of bird migrations (Lack, 1959/63; Lack & Eastwood, 1962; Lack & Parslow, 1962; Parslow, 1962), on the heights at which they occur (Lack, 1960*a*; Nisbet, 1963*a*; Eastwood & Rider, 1965; Gehring, 1967*a*), on their manner of flight (Eastwood & Rider, 1966), on the influence of weather (Lack, 1960*d*; Hassler *et al.* 1963) and especially on the effects of wind on migrants moving over the open sea (Lack, 1958, 1959, 1960*a, b, c*; Drury & Nisbet, 1964; Nisbet & Drury, 1967; Bergman & Donner, 1964), or overland (Bellrose & Graber, 1963; Evans, 1966*a*; Bellrose, 1967*a*). Lack (1962) condensed the light that radar threw on the problems of bird orientation

6

and Evans (1966*b*) has essayed a comparison between the results obtained by visual and by radar observations, while Eastwood (1967) has gathered into book form much of the published data.

We may now attempt to summarize the information garnered by the various methods of field observations, giving preference to radar evidence when there is a conflict. The close linkage between the main migratory movements and the breeding seasons has been fully confirmed. We are not here concerned with the factors governing the latter, summaries having been published by Marshall (1961), Wolfson (1966) and Farner (1967). On the other hand, radar has revealed quite substantial movements throughout the year and the formerly sharp distinctions between the normal (spring and autumn) migrations, 'reversed' migrations and hard-weather movements can no longer be maintained. There follows the implication that navigational faculties cannot be rigidly seasonal.

Even in the normal migratory periods, movements are not spread evenly and the bulk of the migration tends to be concentrated in short periods of a few days. These movements are usually initiated during periods of undisturbed weather with clear skies and light winds. Moderate or strong winds are avoided unless they are favourable. There is no question of the birds simply flying downwind with any wind; they appear to wait for one blowing in the general direction in which they should migrate. There is some evidence of still finer adjustment of headings at the start of the flight, so that the resultant track is in the proper direction.

Once started, a migratory movement may well encounter cloud and unfavourable winds. It may be 'precipitated' over land when it meets the turbulent conditions of a cold front. Over the sea migrants which enter cloud or mist banks become disorientated, milling in all directions and inevitably drifting down wind. There is little direct evidence to support Williamson's (1955) suggestion that in such conditions they descend near the sea, determine the wind direction with reference to the wave patterns and actively fly downwind. Such visual observations as have been made of lowflying birds in mist, rarely migrants (King, 1959), more often foraging seabirds (e.g. Drury, 1959), suggest that they can maintain a heading, probably with reference to the wave pattern, but are completely

7

disorientated when fog closes right down. Well-directed movements continue above fog or low cloud, when ground and sea features are obscured but not astronomical clues. On the other hand there are numerous instances of migration continuing in the normal direction under layer cloud heavy enough to eliminate directional clues afforded by the sun, stars or moon. These instances were over land or at no great distance from the coast; sufficient topographical features would then be available for a point-to-point maintenance of *direction*. Bergman (1964) has shown this to be so in practice. Even at night the ground is sprinkled with a galaxy of lights in densely inhabited areas which would be visible far out to sea from the coasts of (for example) New England. If migration continues for several days in heavily clouded conditions it becomes noticeably less well orientated. Bellrose (1967a) produces rather convincing evidence that airborne birds may be able to determine the wind direction and strength directly, provided that the airflow is turbulent. Vleugel in many papers (e.g. 1959, 1962) has stressed the possibility of wind direction acting as a secondary orientation clue, by day and night.

Over the sea well away from coasts the birds fly on a straight course by day and by night and do not appear to compensate for the effect of changes in wind direction. They may thus drift off their intended track and make landfall far from that indicated by their heading at departure. Thus many of the arrivals of continental birds on the east coasts of Britain can be related to easterly winds acting on a south or SSW movement out of Scandinavia. Occasional observations have been made of wind-drifted night migrants abruptly changing their direction at dawn to a course that offset the drift (Lee, 1963; Myres, 1964). It is not clear yet whether this indicates the operation of a superior navigation by day or a rather stereotyped directional response to finding themselves still over the sea. If migrants coming in off the sea show any reaction to the coast (and they generally proceed straight on) they turn and fly along it in the upwind direction. Radar observations have yet to give unequivocal examples of movements of wind-drifted migrants which have subsequently re-determined the position of their goal. Ringing recoveries certainly indicate that this does occur and there is a little evidence that even juvenile birds on their first migration are so gifted (Evans, 1968).

8

Although the concept of 'standard directions' governing the migratory flights of species, or even local populations, has been fully justified, it is clear that in many cases a single compass course will not take the birds to their winter quarters. Thus warblers and flycatchers travelling SSW to SW across western Europe must at some time swing to east of south to reach their wintering grounds in tropical Africa, while other species which leave to the SE must turn in a clockwise sense (Moreau, 1961). Chaffinches from Norway have been shown to fly a whole series of courses, starting SE, changing at intervals right round to NW. This has the effect to bring them to Britain without crossing the main North Sea (Perdeck, 1962). Such changes of course are not necessarily associated with topographical features (Mook et al. 1957), but the possible influence of experienced birds cannot be ignored. Vleugel (1953) suggested that the increasingly westward trend in autumn could be due to failure to allow for the changing position of sunrise.

To sum up then—if we were restricted to the evidence of field observations, we could certainly conclude that migrant birds, even inexperienced juveniles, at least have the ability to fly on straight compass courses and know which direction is appropriate to the season and the stage of migration. There is evidence that adults have a firm knowledge of the position of the breeding site and probably of the winter quarters and can return to them if displaced by their own efforts or by the wind. Topographical features may temporarily deflect the more low-flying migrants or concentrate them into apparent 'streams'. Landmarks may also serve as reference points for the correction of wind drift or the maintenance of a predetermined direction. Corrections may also derive from an intimate appreciation and assessment of air flow structure. The basic navigational information, however, is most likely derived from astronomical clues. The field evidence thus is that birds are guided almost exclusively by visual stimuli. With this background we may now consider the experimental evidence.

# CHAPTER 2

# Experimental evidence for
# bearing-and-distance navigation

In those species where the young migrate independently of the old we have a natural experiment showing that any tendency to fly in one direction must be part of the bird's innate behaviour; the young Cuckoo provides an extreme example. Where young and old migrate together the former could possibly *learn* the migration direction as well as the final location of the wintering area. This can be tested by holding young birds in the area of their breeding until all others of their species have departed. Rowan (1946) reported briefly that fifty-four Prairie Crows gave 'some 60 %' recoveries of which 'not a single bird had deviated significantly from the standard fall direction'. Schüz (1949) obtained sixteen reports from 247 delayed White Storks, all in the normal migration direction but rather more scattered than usual.

To eliminate the possibility that the unlearned direction was imposed by geographical or meteorological features, Schüz took 144 White Storks from nests in the Baltic region and reared them in West Germany. Released after all local migrants had passed, the tracks followed by the birds (which had colour marks on their plumage as well as leg rings) could be mapped (fig. 1). They show a strong tendency to lead SSE, the appropriate direction for the population from which they were drawn, but quite distinct from that (SW) for the species in the release area.

A more stringent test is to rear and release young birds in areas where the species does not breed. Schüz (1938b) reared 21 White Storks from the Baltic in England; Drost (1955, 1958) 953 Herring Gulls from the Freisian Islands inland in Germany; Bloesch (1956, 1960) 192 White Storks from Algeria in Switzerland; Vaught (1964) 377 Blue-winged Teal ducklings from Minnesota in Missouri. The recoveries indicated migration in the direction normal for the population from which the birds

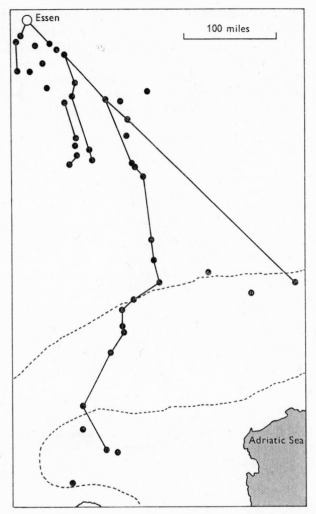

Fig. 1. Reports of young White Storks which were taken from the Baltic and reared and released in Essen after all adult storks had migrated in the autumn. Reports probably referring to the same party of birds are linked together. Note the SSE trend. The Alps are enclosed in the dotted lines. (After Schüz, 1949.)

were taken. The ducks even showed the typical northwards dispersal prior to migration proper. Apart from the release in England, however, the possibility of the birds joining over-flying migrants of their own or related species cannot be excluded. Some other experiments which did not yield much direction information on migration through recoveries, or where the birds were already in their winter quarters, are considered later.

Inherited directional tendencies are indeed readily modified by the example of older birds of the species. This was shown when 754 young White Storks from the Baltic were released in West Germany while stork migration through the area was still in progress (Schüz, 1950). The recoveries, shown in fig. 2, have

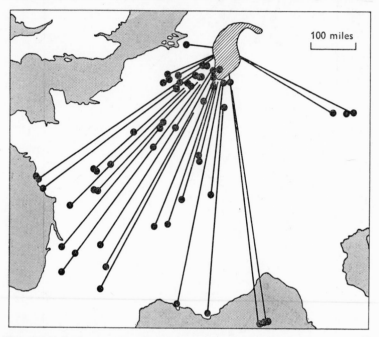

Fig. 2. Reports of young White Storks treated as in fig. 1, but released while adults were still migrating. Note the SW trend. Area of releases is shaded. (After Schüz, 1950.)

a strong tendency to the south-west, approximating to the direction taken by the 'foreign' Storks and contrary to that prevalent in the homeland (compare fig. 1). Similar results appear to have been achieved by Williams & Kalmbach (1943) with 131 Canada Geese and 213 ducks of various species. Migration can even be induced in non-migratory stock by the example of older, migratory individuals of the same species. Thus Valikangas (1933) and Pützig (1938) hatched eggs obtained from English, non-migratory Mallard in Finland and

TABLE 1. *Displacement experiments with autumn passage migrants*

| Species | Total displaced | From | Distance (mls) and direction | Total re-covered | Author |
|---|---|---|---|---|---|
| Starling | 3,013 | Memel | 450 SW | 95 | Krätzig & Schüz (1936) |
| Sparrowhawk | 209 | Heligoland | 410 ESE | 36 | Drost (1938) |
| Hooded Crow | 232 | Rossitten | 360 SW | 54 | Rüppell & Schüz (1948) |
| Starling | 11,247 | Hague | 380 SSE | 354 | Perdeck (1958) |
| Starling | 2,690 | Hague | 760 S | 92 | Perdeck (1964) |
| Chaffinch | 1,265 | Hague | 380 SSE | 3 | Perdeck (1958) |
| Green-winged Teal | 1,500 | Piaam | 450 SSE | 110 | Perdeck (1960) |

the Baltic region. The 116 young were allowed freedom with the local, migratory Mallard and gave nineteen recoveries up to 1500 miles, all within the normal winter range of the local populations. The reverse experiment, rearing migratory stock among local non-migratory stock of the same species has not been reported. However, only six out of ninety-seven Wood Ducks released among a flock of a hundred semi-tame Mallard remained with them over the winter (McCabe, 1947).

Large-scale transplantation experiments require extensive rearing facilities. Another method of investigating directional tendencies has therefore been to capture young migrants that are actually on passage in the autumn and transport them considerable distances to one side of the migration axis, outside the area normally reached by birds passing through the point of capture. Species easy to trap in large numbers have been favoured. Their subsequent movements have been largely pieced together from the ringing recoveries, although plumage marking has been used. Table 1 lists the large-scale experiments that have been reported.

Except for the Teal the same general conclusion was reached. The young birds continued to migrate from the release point in that direction which their congeners follow from the trapping point. The transported birds thus followed a course roughly parallel to the normal and wintered in areas far outside the usual range. Fig. 3 shows in diagrammatic form the results obtained by Perdeck (1958). Although recoveries in the following spring and in subsequent seasons were naturally much fewer, it

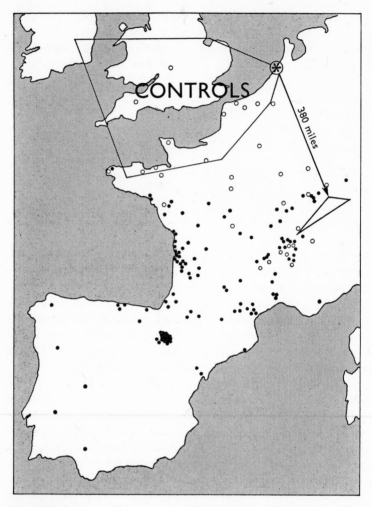

Fig. 3. Recoveries of Starlings displaced from Holland to Switzerland while on autumn migration. Note that the young birds continued to migrate in the direction appropriate to reach the usual wintering area (controls) from Holland. The adults mostly changed direction back towards this area. ○ Adults, ● juveniles. (After Perdeck, 1958.)

has generally been found that the experimental displacement was maintained. The results with Teal were inconclusive, both juveniles and adults were widely scattered around the release point, though the former showed some polarization SW/NE.

Other experiments have involved the displacement of *spring* passage migrants. Drost (1934) moved ninety Starlings and nine Ring Ousels, but obtained only two recoveries. Rüppell (1944) moved 625 Hooded Crows from Rossitten up to 635 miles SW, obtaining 121 recoveries. Again the young birds showed a parallel displacement, breeding outside the normal range and maintaining the displacement in subsequent years.

These displacement experiments with passage migrants required very considerable effort and expense. It is therefore regrettable that their results cannot be taken as *conclusive* proof of the existence of an innate directional tendency in the experimental birds. This is because, for the whole area involved, the general tendency is for migration to be on a NE/SW axis. We have already seen how easily the migration of young birds is influenced by the example of others. There was thus a strong possibility that the transported birds simply joined up with migrants passing the release point and proceeded to *their* winter quarters. While it can be argued that the recoveries of the young Starlings shown in fig. 3 approximate more closely to the Dutch (WSW–SW) than to the Swiss (SW–SSW) pattern, many more recoveries would be required for this to be convincing. The critical test consists of releasing birds in an area where local migration had ceased. Bellrose (1958a) did this by holding 1071 young Blue-winged Teal, intercepted in Illinois, until November, when the species had virtually left the United States (for South America). Half of the young birds recovered the same winter were within 25 miles of the release point but the other fifty-four were almost exclusively in the south-east sector normal for the species. These results would have been even more convincing if the birds had been transported instead of being released near where trapped. Rüppell (1944) added this refinement when he released 271 Hooded Crows caught on spring migration at Rossitten and moved 630 miles SW to a point where local migration had ceased. The subsequent recoveries gave a rather wider scatter than in Rüppell's other experiments, but the tendency was strongly in the north-eastern sector (fig. 4). However, the birds had merely been taken back on their tracks, no real test of their ability to correct for angular displacement.

Another feature of migration displacement experiments is that not only is the original direction maintained but the area of recoveries extends, roughly, for the same *distance* from the release point as it would for control birds released at the trapping point.

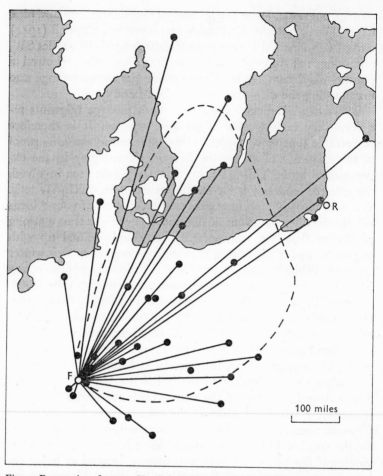

Fig. 4. Recoveries of young Hooded Crows displaced on spring migration from Rossitten (R) to Frankfurt (F) after migration there had ceased. The dotted line encloses an area equivalent to that in which birds released at Rossitten would normally be found. Note that the normal direction and distance of migration is maintained. (After Rüppell, 1944.)

The young birds may finally halt in areas considerably different ecologically from those in which their flight would normally terminate. Contrariwise, Starlings moved from the Netherlands to Barcelona, Spain (Perdeck, 1964, 1967*b*) were released in an area much favoured as a wintering area by the species. Despite this the young birds moved on across the width of Spain. The implication is that the length of the migration journey is controlled by internal factors, that migration continues only while the 'drive' is present. The short-lived nature of the migration impulse is amply confirmed by studies of caged migrants. That migration, once started, proceeds at a fairly steady rate is also confirmed by experimental and observational studies.

The young bird is thus innately equipped at least for the simple procedure of a bearing-and-distance flight to a hitherto unknown area. Provided no unusual circumstances are encountered this enables the untutored population to shuttle back and forth between summer and winter homes. Indeed, it might be considered that this would suffice to carry a bird through all its subsequent migrations. The endpoint of the migration would become a little more precisely defined as the bird became familiar with the topography of its two homes. But in fact the displacement of older birds, with experience of at least one return migration and one season's breeding in the summer home, produced very different results from that of the young birds which we have been considering. There was a strong tendency for the recoveries of these older birds to lie in the direction of, or actually within, the normal winter and summer areas. They ignored the 'standard direction' and the example of local passage migrants at the release point. The difference between young and old birds is best shown in Perdeck's work with Starlings (fig. 3). But tendencies to return to their normal areas were shown by the adult Sparrow Hawks and Hooded Crows. We are thus dealing with a more advanced form of navigation which will be considered in chapter 5 and subsequently.

While the performances of homing Pigeons are also more appropriately considered there, for a long time it has been known that the swiftest and surest returns were obtained with Pigeons carefully trained by releases at gradually increasing distances in *one general direction*. The practice was centuries old when described by Moore (1735). In pigeon racing the overriding aim is a return in the shortest possible time, and, such is the competition, a few seconds at the end of a flight of many hours often make all the difference. As a result pigeon fanciers have adhered strictly to a single line of releases, the race liberation points being as nearly on that line as transport facilities permit. Long sea-crossings form serious obstacles and impose limits in the case of British birds. In America, however, races of up to a thousand miles are regular features.

While a result of this one-direction training could be the building up of a corridor of well known country and a chain of visual landmarks, the later stages involve release at distances of a hundred miles or more from the previous point, precluding a direct view of the latter. Under race conditions (Dinnendahl & Kramer, 1950) the picture is often obscured by more experienced birds being released with the newcomers. But it seemed fairly clear that the Pigeons in fact learnt to fly in that particular direction on release. This was confirmed by Schneider (1906), Kramer & St Paul (1950), Matthews (1951 *b*) and Riper & Kalmbach (1952). These workers showed that after such direction training, Pigeons released in an entirely new direction still flew in the accustomed direction though this no longer took them towards home. Hitchcock (1952) and Walcott & Michener (1967) have followed such misguided flights from the air. Graue (1965 *a*) reports that even one release will impose something of its direction on the next.

The one-direction flights, innate or learnt, that we have been considering thus far have all had an obvious biological relevance in that they coincided with the general directions of autumn and spring migration, or, in the case of Pigeons, with that of home. However, a more puzzling form of one-direction flight has come to light. Griffin & Goldsmith (1955) and Goldsmith & Griffin (1956) released 186 Common Terns taken from nests on the New England coast and on the Great Lakes. The birds showed a distinct tendency to fly off south-east on release, no matter whether the direction of the home colony was NE, ESE, SSW, or W. No plausible reason could be advanced for this conduct. It would have brought the birds to the coast, but this would not benefit those taken from the inland Lakes. Some of the Terns are known to have homed to their nests, but detailed watch was not kept. Then Bellrose (1958*b*) found that Mallard captured in Illinois in autumn and released at several points in various directions flew off very decidedly and consistently NNW. Subsequent recoveries showed that these birds had not maintained this course but had resumed their southward migration. Bellrose suggested that the northward flight was a fright reaction to handling and captivity, the birds turning towards their prairie homes. However Matthews (1961, 1963*a*) obtained similar well marked orientation within a few seconds, to the NW, in a largely non-migratory population of Mallard in SW

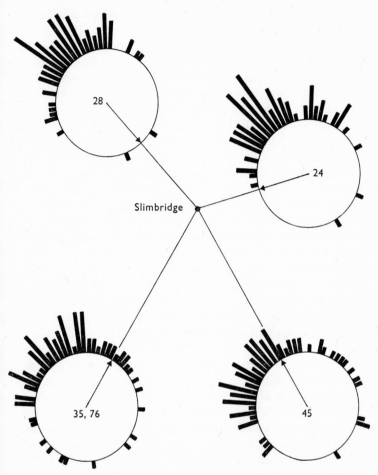

Fig. 5. 'Nonsense' orientation in Mallard. The birds are lost to sight flying predominantly to the north-west no matter in which direction they are displaced. The shortest spoke represents the vanishing bearing of one bird. Distances (in miles) of the release points in the centres. In this and subsequent vanishing point diagrams north is to the top of the page unless otherwise stated.

England (fig. 5). The orientation was consistent no matter in which direction or how far from home they were taken for release, and regardless of sex, age or previous experience; the topography of the release point or the direction of the wind; the time of day or night or the season of the year. Yet the direction was not maintained for more than 20 minutes and most ducks had landed or broken away from the original direction within

19

10 miles (Matthews, 1967). Subsequent recoveries were scattered at random.

A possible explanation of such orientation was that if lone birds flew off in one direction this would facilitate the reformation of the flock. This had to be discarded when it was found (Matthews, 1962) that flocks showed an even sharper orientation in the preferred quarter. Moreover the tendency was apparent when birds were in full view of the place where they were caught and where the rest of the flock remained.

The puzzle became yet more complicated when it was found that Mallard taken from the London Parks also showed a fixed orientation tendency but to the south-east; so did birds in Sweden. Mallard caught near the east coast of England showed NW orientation from August to October, when most of them were local-bred birds. From November onwards, when the migrants from Scandinavia arrived, a catch of birds would give a random pattern, the south-east tendencies of the newcomers apparently offsetting the north-west tendencies of the natives.

Other wildfowl have been shown to have similar definite initial flight tendencies. In Illinois Blue-winged Teal flew S or NW, Pintail WNW, Canada Geese SW (Bellrose, 1963, 1967*b*). Green-winged Teal in the Netherlands and southern France preferred SW, in eastern England NW (Matthews *et al.* 1963).

In such a confusing situation there is justification for calling this behaviour, with no clear relevance to any biological need of the animal, 'nonsense' orientation. This is rather in the sense that Yule (1926) spoke of 'nonsense' correlations, sometimes yielded by time series without causal connections being apparent. While such behaviour is nonsensical to us we must allow that it may make sense to the bird. It has also, as we shall see (p. 78), made nonsense of some interpretations of bird navigation based on the performance of homing Pigeons.

Though in some cases the orientation is one that would take birds disturbed on their feeding grounds towards a likely refuge, in others it does not. We are probably not dealing with a simple escape tendency learned in relation to the local topography. It is certainly shown at very short distances, 1 to $2\frac{1}{2}$ miles, among landmarks which must be known to the birds (Matthews, 1962 & in preparation). Then migrants apparently carry the 'nonsense' with them and even if a SE tendency is appropriate in Sweden, it could not be in eastern England, where the natives prefer NW. There is evidence that the tendencies are innate. Mallard reared to flying in large aviaries in SW England showed a northerly tendency though their uncertain flight reduced the clarity of the result (Matthews, 1963 a). Pintails reared by Hamilton (1962 b) in a flight cage in Manitoba gave a definite orientation to the NW on their first release. This was also the direction of home, but in the absence of releases from other directions, and in view of Bellrose's results with wild-caught Pintail, this was probably another instance of 'nonsense' orientation. Certainly Heyland (1965) found such an orientation in Pintail, and in Mallard, trapped in early autumn in Saskatchewan.

Another instance of a fixed orientation in birds has been demonstrated by Emlen & Penney (1964) and Penney & Emlen (1967) using Adelie Penguins. These were taken from their nests on the edge of Antarctica to release points on featureless snow-ice plateaux in the interior, and the directions in which they walked off were plotted. The departure patterns were orientated northwards (NNE for birds from one colony, NW from another) rather than directly towards the colonies. The same orientation was also shown in birds released to the north of the colony, further demonstrating its 'nonsense' nature. In the Antarctic a northward tendency would, at the Pole, where some birds were released, be rather inevitable. It would however clearly be useful in that it would take the birds to the coast wherever they happened to be released.

Certain riparian arthropods also have fixed orientation tendencies, but in most cases the direction taken also has a clear relevance to the lie of the shore line on which the creature lives, e.g. the sand-hopper *Talitrus* (Papi & Pardi, 1953). In other cases, however, such as the pond-skater *Velia*, a constant direction, south, is taken up by populations in many parts of Europe (Birukow, 1956; Emeis, 1959). Again, the wolf-spider *Arctosa* has local populations with various escape tendencies which are quickly learned in relation to the slope of the bank on which they are reared (Papi & Tongiorgi, 1963); but when reared in flat, uniform surroundings a northward tendency develops gradually but spontaneously. So a simple ecological answer is not available even in these arthropods. Nevertheless their orientations, and likewise the 'nonsense' orientations of birds, are suitable for investigating *how* one-direction movements are undertaken even if it is not clear *why*.

# CHAPTER 3

# The physical bases of diurnal one-direction navigation

Displacing migrants and awaiting the accumulation of re-coveries is too cumbersome a technique to study the means whereby birds determine and maintain their one-direction flights. A most important advance was thus achieved by Kramer (1949) when he found that caged passerines manifested direc-tional tendencies during the periods of intense activity that coincide with their migration times. The birds tended to head in a particular direction whether they were hopping about or sitting on a perch fluttering their wings. The birds performed well in a drum-shaped cage, less than a metre in diameter, which has since been widely used and appropriately become known as a Kramer-cage. An observer lying below the cage estimated the general heading of the bird over a short period. The strength of the orientation was then represented by the con-centration, or otherwise, of such headings in certain sectors. It was, however, difficult for the observer to be completely objective and to decide when and what directed behaviour is being shown, particularly with a rapidly moving bird. It is also a very weari-some procedure.

Merkel & Fromme (1958) and Mewaldt & Rose (1960) there-fore introduced automatic registration of the bird's position on the circular perch. The ring was broken into a dozen sprung perches, pressure on which was recorded by an electrical contact operating a marker on a moving tape recorder. The apparatus has the basic weakness that the bird's position in the cage does not necessarily reflect the orientation of its body. Aagaard & Wolfson (1962) developed a variant in which the bird was free to move from a large circular cage into any of eight small cages attached to it. A more definite 'decision' was thus re-quired of the bird and its movements within the small cages were picked up by a magnetic microphone. Hamilton (1966) has preferred to record the visual image of the bird by an elaborate system based on time-lapse infra-red photography and requiring computerized analysis. Such complicated systems can absorb far too much of the experimenter's time in simply getting and keeping it working. Indeed neither have resulted in much published data. A welcome return to simplicity was therefore achieved by the Emlens' (1966) ingenious device. The bird is made to stand on a pad, impregnated with indelible ink, in the centre of an inverted cone of absorbent paper. Scrambling up and sliding back the bird left a permanent record (which can be judged on a density standard) of its preferred directions of escape.

The limitation of experimental investigation to the season and periods of migratory restlessness is inconvenient. Kramer (1952) therefore devised a technique whereby birds were trained to take food from one container, situated in the appropriate direction, of a circle of twelve. As the bird had to commit itself to a definite choice this could readily be recorded automatically if required.

Although some anomalous tendencies have been found, a number of workers have reported orientation in caged birds in the daytime which approximates to the migratory direction appropriate to the season. The species involved were Red-backed Shrike, Blackcap and Starling (Kramer, 1949, 1951 a), Starling (Perdeck, 1957), Greenfinch, Goldfinch, Linnet, Siskin, Chaffinch, Brambling, Crossbill, Parrot Crossbill, Hawfinch, Scarlet Grosbeak, Barred Warbler, Garden Warbler, Blackcap, Reed Bunting, White Wagtail, Meadow Pipit and Wood Warbler (Shumakov, 1965). Jacobs (1962) found general but not detailed correlation with seasonal directional changes in Chaffinches.

Migrational (and also trained) directional tendencies were shown with landmarks hidden from the bird's sight, by means of a shield round the cage or by placing the latter in a pavilion with symmetrically arranged windows. The birds therefore did not need landmarks to determine the direction, but they were very prone to align their directions with reference to them, even to minute irregularities of the shielding walls. It was therefore a very necessary feature of the apparatus that the latter should be capable of rotation at intervals so that marks did not remain in any one direction for long.

When the sky as well as landmarks were excluded and experiments made in closed rooms, no orientation natural or learned could be obtained by Dijkgraaf (1946), Kramer (1951 a), Sauer (1957), Shumakov (1965), Walraff (1966 a) or Emlen (1967 b). The claims of Merkel & Fromme (1958) and Fromme (1961) to have demonstrated migration orientation inside tents and buildings therefore runs counter to the concensus. Moreover, when carefully examined, their evidence is not convincing. The automatically recorded movements of their birds (Robins and Whitethroats) were calculated as deviations from the value expected if activity were equally divided between eight perches. Only the immediate zone on either side of the expected line was shown on their graphs. Deviations of as little as 10 to 20 % from the chance value were thus visually exaggerated. A more objective presentation as a spoke diagram was unconvincing as evidence for orientation. It is not clear why, moreover, certain nights of so-called 'reversed migration' were omitted from the calculations. At most there might be progressively poorer 'residual' orientation as the birds are moved farther from the open air situation (where they had been caged for several months prior to the experiments). Carry-over of orientation by transfer to marks within the cage had been noted by Kramer (1952), Mewaldt et al. (1964) and Shumakov (1965). Ambient sounds could serve in the same way. It is suggestive in this connection that any trace of orientation was eliminated by shutting the doors of a steel chamber in which the birds had been placed. Small deviations from uniformity in the surroundings can be of critical importance when dealing with marginal evidence for orientation. Thus the birds avoided one perch against a wall of the testing cage which was 2·8° C warmer than the others. Merkel et al. (1963) and Merkel & Wiltschko (1966) have sought to refute criticisms of their analysis and published other, no more striking, results.

More satisfying than rejection on the grounds of flaws of presentation or experimental technique, are further careful experiments with the same species giving negative results. These have been presented by Perdeck (1963). He examined the migration activity of Robins automatically recorded in circular cages (which were turned frequently) situated inside a wooden building. Some suspicions of orientation were found while other Robins were calling in the same building, but when these were removed the birds under test showed no orientation whatsoever. When the cages were *not* turned during the experiment the rapidly acquired preferences of birds for certain perches gave a deviation from a statistically chance distribution but without a significant accumulation in any one general *direction*.

The fact that Fromme (1961) claimed disorientation only in a steel chamber raises the possibility that the effect was due to interference with the earth's magnetic field. Middendorf (1855) first suggested that birds were capable of detecting the magnetic poles and of maintaining their bearing therefrom. Similar ideas have continued to crop up at intervals. Cathelin (e.g. 1935) proposed that birds followed some mysterious electro-magnetic currents. Fromme, however, found there was no gross deflection of a compass needle in his chamber and that a powerful artificial magnetic field did not influence the orientation of his Robins. This was in agreement with the negative results obtained by Kramer (1949) with caged migrants, and by Rochon-Duvigneaud & Maurain (1923), Griffin (1944, 1952c), Henderson (1948), Wilkinson (1949) and Orgel & Smith (1954) who all failed to detect any sensitivity to intense magnetic fields (static or moving) in birds. Meyer (1966b) deliberately used fields that did not differ so markedly from the normal and tested Pigeons by operant conditioning. Again no sensitivity was shown. By contrast Eldarov & Kholodov (1964) claim that exposure to a constant magnetic field of from 0·16 to 1·7 ergs increased the amplitude and changed the character of the motor activity of caged passerines. Recently Merkel & Wiltschko (1965) and Wiltschko & Merkel (1966) have claimed an effect on the orientation of caged migrants by imposed magnetic fields. But again the statistical analysis is unconvincing and the effect, at best, very marginal. Pigeons orientating their flight in a trained direction have done so undisturbed by powerful magnets swing-below the head (Matthews, 1951b) or attached to the wings (Riper & Kalmbach, 1952). North-west-seeking Mallard were also quite unaffected by swinging neck-magnets (Matthews, in preparation).

Although we can probably discard the idea that birds orientate with regards the earth's magnetic field (the even more remote possibility that they can find their postion thereby is considered later, p. 104) this does not necessarily apply to all animals. Distinct physiological effects are known to be produced

25

by a varying magnetic field. Barrett (1883) claimed that certain human 'sensitives' could detect magnets by their luminosity. More recently Becker (1963) has reviewed studies of the so-called magnetic 'phosphene' effect, luminous areas induced in the field of vision by magnetic pulses (or by pressing on the eyeballs), and of other reported biological effects. Lissmann (1958) has demonstrated a remarkable form of orientation over very short distances in certain fish. These set up a weak electrical field around themselves and apparently detect changes in their surroundings by changes in impedance. Such fish certainly react to a moving magnet, but not, apparently, to the earth's field (Lissman & Machin, 1958). Brown and his co-workers in an avalanche of papers (e.g. Barnwell & Brown, 1964) have claimed orientational responses to an artificial magnetic field by snails, by planarian worms and protozoa. Schneider (e.g. 1963) makes similar claims for cockchafers, Becker (1963) for termites and Becker & Speck (1964) for flies. Kimm (1960) however rejects the possibility of reaction to the earth's magnetic field on the grounds that any possible force produced in the animal would be too small to be detected—as Wilkinson (1949) had earlier argued for birds.

Martorelli (1907) and Geyr (1922) suggested that the NE–SW trend of migration in western Europe was caused by birds moving N/S and being deflected passively by the Coriolis force in the same way as air masses. However, there would be no appreciable effect on the comparatively short flights of typical migrants, and we now know that the NE–SW tendencies are present in the caged migrant. The hypothesis of Roberts (1942) and Beecher (1951) required birds to determine the direction of the earth's rotation, to serve as a compass guide, by actual detection and measurement of the Coriolis force. The extreme improbability of such achievements is discussed later (p. 102).

Landsberg (1948) and Suffern (1949) suggested that regular migrations, particularly those of oceanic species, could be accomplished by flying continuously downwind, 'pressure pattern flying' in aeronautical jargon. While there is evidence that wind plays an important secondary role in migration and that birds may wait for a favourable wind (p. 7), there is none that it serves as a primary orientation factor. Again, caged birds shielded from the wind show orientation.

We may conclude that birds do not orientate with reference

to any unknown, magnetic, rotational or pressure stimuli, and suggest that visual stimuli are of paramount importance. Kramer (1951 a) found that a Starling placed in his circular cage apparatus under a heavily overcast sky was completely disorientated. When the cloud dispersed the bird quickly became strongly orientated. Although the direction selected was not that appropriate to spring migration, to the north-east, its NW tendency can be considered as an early instance of 'nonsense' orientation. Other tests showed orientation, somewhat less precise, under cloud cover of a moderate degree of compactness, or when the light was diffused artificially by semi-transparent paper over the windows.

Perdeck (1957) confirmed the disorientating effect of heavy overcast on caged day migrants. One of his Starlings picked up its orientation over several days of cloudy tests, but it may well have made use of acoustical reference points. Shumakov (1965) found that under unbroken cloud daytime directions were random, determined by cage features, or were not manifested at all. He showed that Chaffinches and Barred Warblers could maintain a direction for some hours after the onset of heavy clouds if, when the sun was out, landmarks were also visible. Kramer and his colleagues showed that learned orientations of Starlings and Pigeons likewise broke down with overcast, as did those of Pintail and Blue-winged Teal ducklings (Hamilton, 1962 a).

Returning to free-flying birds, the orientation of Pigeons in a training direction (Matthews, 1951 b) and likewise 'nonsense' orientation of Common Terns, Mallard, Pintail, Blue-winged Teal, Canada Geese were all hampered by overcast (Griffin & Goldsmith, 1955; Bellrose, 1963; Matthews, 1963 c) as was that of the free-walking Adelie Penguins of Emlen & Penney (1964).

The clouds must be really dense for complete disorientation, suggesting that a simple, compass orientation is picked up as long as the approximate position of the *sun* can be discerned. The exact limits of cloud density can be explored by measuring brightness round from the calculated sun position by means of a spot-photometer (Matthews, unpublished). Frisch *et al.* (1960) used a refined photographic technique to demonstrate that bees could localize the sun through cloud, probably by a sensitivity to ultraviolet light (which is unlikely in birds).

Kramer proved that his caged Starlings needed to see the sun or the sky within 30 to 45° round it; Hamilton's ducklings were

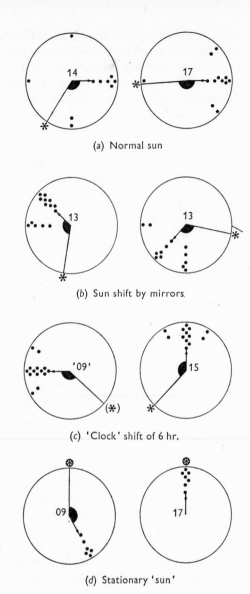

(a) Normal sun

(b) Sun shift by mirrors.

(c) 'Clock' shift of 6 hr.

(d) Stationary 'sun'

Fig. 6. Analysis of sun-compass orientation in caged Starlings. (a) The birds select the same direction at different times of day. (b) The orientation is changed by displacing the sun's apparent position by mirrors. (c) The orientation can equally be changed by shifting the bird's internal 'clock'. (d) The angle taken up with reference to a stationary 'sun' changes through the day. Time is shown in hours in the centres. Dots represent activity periods (a, b) or food choices (c, d). (Derived from Kramer (1952) and Hoffman (1954).)

confused when the sun was below the horizon. Their birds thus showed no ability to orientate with reference to the pattern of polarised light coming from blue sky well away from the sun's position, as can many arthropods. Montgomery & Heinemann (1952), in laboratory training tests, also failed to demonstrate in Pigeons any special sensitivity to polarised light.

When Kramer placed mirrors at the windows of his pavilion in such a way that, viewed from within, the sun's apparent position was changed through 90° either way, the Starling's 'nonsense' orientation was deflected accordingly (fig. 6b). Similar mirror experiments were done with direction-trained Starlings (Kramer, 1952) and Pigeons (Kramer & Reise, 1952), also with positive results.

With free-flying birds it is difficult to shift the sun position at will. However Matthews (1963c) reports one lucky natural experiment. In overcast conditions the sun was setting, completely hidden, in the south-west; a break in the clouds appeared low in the north-west and flushed red, looking to human eyes just as if the sun were setting there. Mallard released in this situation flew north-*east* instead of north-west, as had controls released when the true sun position was discernible on the same day at the same place.

The sun itself is thus clearly the reference clue used. Schneider (1906) had suggested that trained Pigeons might fly at a fixed angle to the sun position, being always released in the same direction and at about the same time of day. However, after training in which these two factors, direction and time of release were kept constant, Matthews (1951b) found that Pigeons would still orientate well in the home/training direction when released 6 hours earlier or later than normal. Again, Kramer's caged birds would take up their spontaneous orientation at different times of day (fig. 6a). When caged Starlings, Pigeons (Kramer & Reise, 1952) and Western Meadow Larks (St Paul, 1956) were trained to feed in one direction at one time of day and then tested at another time, most of them still took up the original direction. Mallard flew off in the same 'nonsense' direction from sunrise to sunset (Matthews, 1961). All these birds would appear to be adjusting the angle they made to the sun's position according to the time of day.

The situation was examined further (Kramer, 1952, 1953c) by using an artificial sun, a light source subtending an angle corresponding to that of the sun's disc. It could be raised or lowered vertically and thus set at an altitude corresponding to that of the true sun at a particular time of day. It remained fixed in the horizontal plane. Starlings could be trained to accept the artificial sun in lieu of the true sun and to take up a given direction, say 'west', with reference to it and at particular times. If the bird was now tested at a different time of day it still orientated to the 'west' by making a different angle to the 'sun', i.e. it reacted as if the 'sun' had a diurnal movement (fig. 6d).

In most species tested there were individuals which did *not* compensate for the sun's movement but took up a fixed angle to the sun position, i.e. a changing compass direction. It is not really clear whether these particular birds were genuinely ill-equipped or whether they reacted in this limited way because of the experimental conditions. In most cases it was actually much more difficult to train a bird to go at a fixed angle even to an artificial light source. Thus Matthews (1952b) achieved this only by first marking the food pot (in a circle of eight) with a landmark, and then successively diminishing this in size while the Pigeons slowly learned the fixed-angle concept.

It is probable that learning is normally involved in the establishment of a complete time-compensating orientation. Thus Hoffmann (1954) took six 12-day-old Starlings from their nest box and reared them without any direct view of the natural sun. He then direction-trained them at one time of day with an artificial sun. Only two, tested at other times, compensated as for a sun moving clockwise and then with gross errors in terms of sun speed. The others never did more than maintain the original training angle. Fischer (1961) has shown that angle-constant animals (lizards in this case) can be converted into direction-constant ones by training at several times of day. They then take up the correct direction at intermediate times, by a process of, as it were, extrapolation.

Implicit in the progressive changes of sun-angle made by the birds is their possession of some form of time-keeping mechanism. Such 'biological clocks' have been the subject of intensive investigation in recent years and reference may be made to summaries in Bünning (1964), Harker (1964), and Sollberger (1965) Many of the advances have been through orientation studies themselves, the creatures acting, as it were, as animate clock

30

hands. Here we shall concern ourselves with those aspects of the problem which have a direct relevance to navigation. While much work has been done on birds, some of the critical questions have only been examined in the case of other animals, and these have to be quoted *faute de mieux*.

To summarize: there is general agreement on the existence of intrinsic rhythmic time-keeping processes, whether these be neural or hormonal in nature, which serve as the 'clocks'. It also appears that the rhythms can continue in the absence of external changes but that they must be initiated by an external stimulus. Moreover, in natural conditions, the rhythms are kept in step with the external rhythmic events (derived ultimately from the earth's rotation), by pacemakers (*Zeitgeber*). Of these the light/dark succession of day and night is one of the most important.

By shifting the light/dark cycle artificially it is possible to reset the internal clock and so afford another opportunity of checking the reality of sun orientation. Thus Hoffmann (1954) trained two Starlings in one direction with reference to the natural sun, and then kept them for 12 to 18 days in an artificial day retarded 6 hours from normal. They were then tested under the natural sun at times when its height was the same as would be 'expected' in the artificial day conditions, 09·00 artificial time and 15·00 normal time. He found that their orientation changed through 90° clockwise from the training direction, i.e. roughly 15° per hour (fig. 6c). This new orientation was maintained when the birds were kept under constant light conditions for 23 days. It was then reinforced by training under the retarded day conditions for 29 days. Finally after 12 to 16 days under normal light conditions in an outside aviary the birds were found to have returned to their original orientation, i.e. had regained 6 hours. Schmidt-Koenig (1958) direction-trained Pigeons and then shifted their clocks by 6 hours, fast and slow, and by 12 hours (in summer, when the real and artificial days overlap). He found the predicted deviations of choice through 90° left and right, and through 180°.

'Nonsense' orientation of free-flying birds has also been changed by subjecting them to shifted artificial days. Thus Matthews (1963c) duplicated with free-flying Mallard the results with trained caged birds, his birds going respectively SW, NE (fig. 7) and SE instead of in their 'nonsense' direction, NW.

31

Probably, as we shall see later (p. 140), Schmidt-Koenig's (1961 a) experiments with free-flying Pigeons should be included under this heading.

A neat converse test would be to shift birds in longitude and so subject them to an external time change without shifting their internal 'clocks'. In temperate latitudes this is difficult because of the great distances involved, about 4000 miles to produce the equivalent of a 6 hour shift. Emlen & Penney (1964) and Penney & Emlen (1967) however took advantage of the convergence of the lines of longitude at high latitudes (*ca.* 80° S)

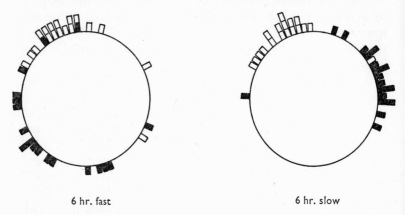

6 hr. fast                    6 hr. slow

Fig. 7. Time shifts of their internal 'clocks' change the 'nonsense' orientation of free-flying Mallard by day, showing it to be based on a sun-compass. (From Matthews, 1963*c*.)

where a displacement of 900 miles resulted in a longitude shift of 120°, equivalent of an eight hour time shift. Using Adelie Penguins showing a northward 'nonsense' orientation they found the birds deviated anti-clockwise when transported east (internal clock slow) and clockwise when transported west (internal clock fast). This would be expected from a time-compensated sun-orientation mechanism operating south of the equator. The amount of deviation observed was, however, rather less than would be expected from an equivalent time shift. Birds held for three weeks apparently shifted their 'clocks' to the local time despite the small fluctuation in light there is in the 24 hour Antarctic day.

If time-shifted birds are released when the time and the

artificial days do *not* coincide, in their subjective night, their orientation behaviour will give us clues as to the way in which the sun-orientation mechanism works during the night. The problem could be considered (for northern latitudes) in terms such as 'where do the birds ''think'' the sun goes at night, on through the north or back through the south'. If the former, then the 'expected' sun position will be opposite the actual sun throughout the 'night' of 12 hour shifted birds and direction of flight would be, e.g. SE instead of NW, no matter what time of the actual day the birds are released. In the second case the 'expected' position will coincide with the actual sun position at 'midnight'/midday and a regular change of direction will occur through the day, e.g. Mallard normally flying NW should (at the equinoxes) go SE at sunrise, SW at 09.00, NW at midday, NE at 15.00, and back to SE at 18.00. Tests with several hundred Mallard showed that they conformed with this 'sun-back' model (Matthews, 1963 *c* & in preparation). We would probably do better to consider the 'sun-back' behaviour as representing a nocturnal 'unwinding' of the sun-angle correcting mechanism. Thus if we consider fig. 8, a bird flying NW needs to make a progressively smaller angle to the sun position during the day. The small sunset angle must be returned to the large sunrise angle in the course of the night. This could take place by a reversal of the daytime process, by a rhythmic oscillation. The partially 'unwound' angle-correction would not normally be functional, but, when the sun suddenly appears, the mechanism 'locks on' and so orientates the animal.

Schmidt-Koenig (1961 *c*) tested two phase-shifted Pigeons in their physiological night in a circular choice apparatus. They behaved as if they went through an unwinding process. Under laboratory conditions the state of the angle-correcting mechanism can be examined by noting the angle taken up to an artificial sun at different times during the night of an animal *not* subjected to a time shift. In this way a variety of other animals have been shown to have 'unwinding' sun-orientation mechanisms, the amphipods *Talitrus* (Pardi, 1954) and *Orchestia* (Debenedetti, 1962), the pond skater *Velia* (Birukow, 1956), the beetle *Phaleria* (Pardi, 1958), the spider *Arctosa* (Tongiorgi, 1959), and the tropical fish *Crenicichla* (Braemer & Schwassmann, 1963). However, the same workers have obtained contradictory results either using the same animals under different

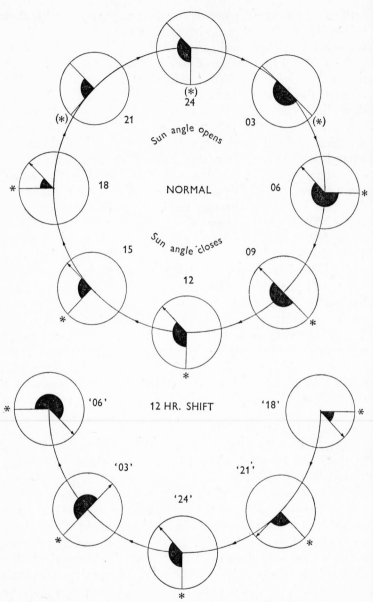

Fig. 8. The oscillation hypothesis of the sun-compass. During the day (06 to 18 hours) the angle set to the sun decreases. During the night the mechanism opens out the angle (as if the sun were running backward through the south) until the dawn setting is regained. Birds with their clocks 12 hours out of phase will make the errors indicated in the lower semicircle if released in their 'physiological night'.

conditions, e.g. *Velia* by Birukow & Busch (1957) or by using related animals under the same conditions, e.g. the temperate fish, *Lepomis* and *Aequidens* by Braemer (1959). In these cases, as also with bees, *Apis* (Lindauer, 1957) and lizards, *Lacerta* (Fischer, 1961) the animals' angle-correcting mechanism apparently 'wound-on' through the night, through the full 360° in 24 hours. Schwassmann (1960) has suggested that the apparent existence of two methods of resetting the angle-correcting mechanism might reflect the evolutionary origin of the animals concerned, those with 'winding-on' mechanisms originating in high latitudes, where the sun is seen to make a complete circuit, those with 'unwinding' mechanisms coming from tropical zones (though this would not explain the existence of two methods in one species, *Velia*).

So far no experiments have been done with bird populations indigenous to the Arctic. Hoffmann (1959*b*) took caged, direction-trained Starlings from Germany in summer to a point 68° N and continued their training there. He found that they could orientate, albeit somewhat inaccurately, while the sun passed through the northern half of the sky. However the birds had had several weeks of experience of the continuous day before the critical tests were made. They may have adjusted to it by a learning process. Schmidt-Koenig (1963*d*) direction-trained Pigeons at 36° N and then tested them directly at 71° N. At night the birds sometimes treated the sun as if it moved clockwise, sometimes anti-clockwise. Papi & Syrjämäki (1963) made the more precise experiment of taking spiders, *Arctosa*, from Italy to Finland, at 69° N, and keeping them on the Italian day/night schedule until the tests were made. At midday the spiders were then well orientated, at midnight quite disorientated. Yet Finnish populations of the *same* species were well orientated at both times.

A further fascinating field of experiment would involve the transportation of birds from northern latitudes across the equator into southern latitudes. There they would find the midday sun in the north instead of the south and the apparent movement of the sun would be anti-clockwise, to an observer facing it, instead of clockwise. Then a transported bird seeking to go NW would do so correctly at sunrise but go SW at 09.00, SE at noon and NE at 15.00, and again correctly NW at sunset (cf. the directions taken in the subjective night of a time-shifted bird, p. 33).

35

Schmidt-Koenig (1963 b) has exploited this experimental opportunity with birds. Five Pigeons were direction-trained in a circular choice apparatus in North Carolina (36° N) and then transported to and tested in Brazil (1½° S) and Uruguay (35° S). The birds' choices were very widely scattered in the new situations, but nevertheless they could be said to react as if they were still north of the equator and so made the expected directional 'errors'. The birds were only exposed to the sun during the short test period; it remains an open question whether they could eventually adjust to the new conditions. Hoffmann (personal communication) has found the expected direction reversals in trained Starlings transported across the equator. More experiments have been done with other animals and may again be cited as supporting evidence. Kalmus (1956) and Lindauer (1957, 1959) moved bees across the equator (or at least across the latitude where the noon sun was vertically overhead) and Hasler & Schwassmann (1960) and Braemer & Schwassmann (1963) carried out transequatorial shifts with fishes. The results were either a disorientation or false orientation in the predicted directions. In the case of bees the weight of the evidence indicated that at least the generation descended from translocated queens was able to make correct allowance for the local sun movement. With fish the adaptability seems to vary with the latitude of origin. Tropical species, which normally experience a change in direction of sun movement at each equinox, responded more readily to a change in sun movement resulting from transportation than did those from higher latitudes.

Because good compass orientation has been achieved when the sun (real or artificial) is not at the appropriate height for the time of day (real or subjective), several authors from Kramer onwards have concluded that the animals are concerned only with the horizontal component of the sun's position, its azimuth, and not at all with the vertical component, its altitude. At first sight a simplification, this interpretation does lead to complications. Consider fig. 9, which represents the relation between sun and observer in Ptolemaic terms, i.e. the sun moves and not the earth. From this it will be appreciated that the horizontal (azimuth) component of the sun's movement is small early in the day and large around noon. To take a concrete example let us consider a bird in 51° lat. (southern England) taking up a particular direction on 22 June. Suppose it does so at 06·00 hours local time, and then 4 hours later at 10.00 hours has to do so again. The original bearing with reference to the sun's position in azimuth will have altered through 53°. To take up the same direction after a further lapse of 4 hours, at 14·00 hours, will require a further alteration not of 53° but of 102°, nearly twice as great. Nor does this diurnal variation in the rate of change of azimuth remain constant. From its position on 22 June (the summer solstice) the sun's arc sinks below the horizon until an extreme position is reached on 22 December (the winter

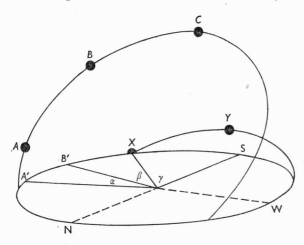

Fig. 9. Perspective diagram of the sun arc in summer (upper) and winter (lower) solstices for 51° N. The sun takes the same time to move from *A* to *B* as from *B* to *C*, but its downward projection moves round the horizon at very different rates from *A'* to *B'* and *B'* to *S*. Again, *BC* and *XY* are the same length of sun arc traversed in the same time, but the equivalent movement round the horizon (azimuth change) *B'* to *S* and *X* to *S* is very different.

solstice). This means that the same amount of movement around the sun arc now subtends a smaller angle around the horizon. Thus the bird correcting its bearing between 10·00 and 14·00 hours will now have to make a change of only 56° instead of 102°.

At lower latitudes the situation departs even more radically from the ideal one (at the poles) of the sun having a constant rate of change in azimuth, of 15° per hour throughout the 24 hours, to complete the circle. On the equator, at the equinox for instance, the sun does not change its azimuth (090°) at all from sunrise to noon, then changes abruptly through 180°, and maintains an azimuth of 270° until sunset. Of course the sun actually proceeds smoothly in a vertical arc passing directly overhead. The zig-zag is purely an artifact obtained if we plot azimuth against time on a graph (fig. 10).

Now if it could be shown that an experimental animal changes its angle to the sun *always* at 15° per hour, we would be justified in concluding that it could only interpret (very roughly) the horizontal angle. If it showed varying rates of change according to time and season this could mean *either* that it made the necessary corrections by an innate 'nautical almanac' *or* that information was being derived from the sun's altitude whether directly, or indirectly through changes in day length.

Many of the detailed experiments with birds and with bees (e.g. Renner, 1959) were done at seasons and latitudes such that divergences from the 15° per hour angle-correction could not be detected against the scatter of the experimental results. Papi *et al.* (1957), using a spider, *Arctosa*, were the first to demonstrate a varying rate of change of angle to the sun position according to the time of day. Braemer & Schwassmann (1963) provided unequivocal evidence that for equatorial fish, trained in those regions, the rate of change was slower in the morning and evening than it was around noon. They have also proved that the sun height is a factor governing this change by showing the sun to the fish in a mirror. If the mirror was tilted so that the reflected sun appeared higher (by 22·5°) the fish responded by swimming at a larger horizontal angle, as if it were later in the day. Schwassmann & Hasler (1963) by transportation of trained fish to different latitudes confirmed that the sun altitude was both responded to and affected the horizontal orientation. These conclusions accorded with earlier work on fish (in a temperate latitude) by Braemer (1960). Here time-switched fish showed, in their subjective night, the expected false orientation except when the sun was at its highest point (noon). They then showed conflict between the true and false directions and also a form of 'displacement behaviour', swimming in small circles and lying on their sides.

Another instance where a change of orientation behaviour is directly consequential on the sun's altitude concerns the experiments already cited with Pigeons, bees, fishes and lizards, in which the direction-trained animals were tested when the sun was near (3 to 11°) the zenith (i.e. directly overhead). In all cases the animals appeared to be disorientated.

Yet another factor correlated with sun altitude is the length of day. By increasing the length of the (artificial) day Birukow & Busch (1957) could increase the amplitude of the oscillation left and right of 'sun' position that *Velia* would make during the day. That is, in seeking south after midday the animals would correct further to the right of the 'sun' before beginning to

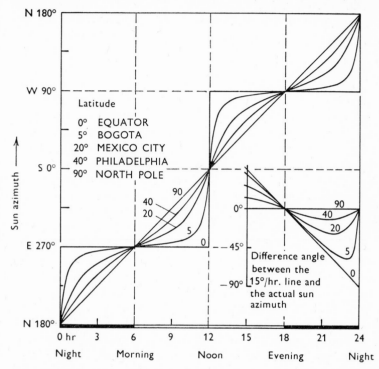

Fig. 10. Sun azimuth as a function of local time for different latitudes at the equinox. The small curves at the right indicate the errors which will be made if correction for sun movement is made at a constant rate of 15°/hour. (From Braemer, 1960.)

'wind-back' through the night (p. 33). Schwassmann & Braemer (1961) also showed that fish subjected to artificially lengthened days increased the velocity of angular correction, while those subjected to shortened days decreased it. Mittelstaedt (1962) has formalized the sun-compass orientation of insects, in terms of his bi-component hypothesis whereby each necessary input is made in terms of its sine and its cosine value.

Although comparable results have not yet been reported in the case of birds it is probable that similar adjusting mechanisms will be found both in respect to season and to hour of day, i.e. that the birds are *not* arbitrarily considering only one component of the sun's movement; that azimuth-only correction is an artifact consequent on relatively crude experimental technique.

# CHAPTER 4

# The physical bases of nocturnal one-direction navigation

Kramer's earliest orientation studies (1949) were with caged *nocturnal* migrants, Red-backed Shrikes, Blackcaps and White-throats. These did show directional tendencies at night but only when they were exposed at the test site at or before sunset. There were indications that they mistook the sky-glow over a city for the sunset, leading to false orientations. St Paul (1953) then showed that the Red-backed Shrike and another typical night migrant, the Barred Warbler, had a well developed time-compensated sun-orientation mechanism. The last eight species listed under Shumakov (1965) on p. 23 are likewise night migrants showing sun orientation. For a while, therefore, the general opinion was that night migrants determined their direction of flight around sunset and then maintained it as best they could through the hours of darkness. The landscape below and perhaps the general moon position and star pattern might then serve in a secondary role to maintain the orientation.

However, the picture was radically changed by Sauer & Sauer (1955). They exposed Blackcaps and Garden Warblers in a Kramer-cage in the middle of an autumn night. After some confusion these took up the direction appropriate to the local population of their species—and these were hand-reared birds which had never been allowed to see the natural sky before (fig. 11). The same individuals tested in the spring (Sauer, 1957) showed the appropriate reversal of their directional trend. Sauer & Sauer (1959) found strong northerly tendencies at night in one Whitethroat, one Garden Warbler, four Wood Warblers and a Lesser Grey Shrike caught and tested in SW Africa at the time of their return migration.

Nocturnal orientation appropriate to the season in caged migrants has been confirmed in California by Mewaldt & Rose (1960) and Mewaldt *et al.* (1964) using White-crowned Sparrows;

by Hamilton (1962*d*, 1966) using Bobolinks, Yellow-billed Cuckoos and Black-billed Cuckoos; by Shumakov (1965) with Ortolan Buntings, Tree Pipits, Red-backed Shrikes, Grass-hopper Warblers, Robins, Thrush-Nightingales, Dunnocks, Scarlet Grosbeaks, Barred Warblers, Garden Warblers and Blackcaps; and by Emlen (1967*b*) with Indigo Buntings and (1967*a*) Rosy Grosbeaks. Hamilton (1962*a*) found that duck-

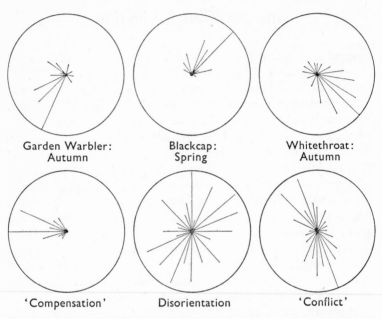

Fig. 11. Orientation of the migratory activity of caged warblers under the dome of a planetarium. Top row, with normal settings of the star pattern; bottom row, after experimental shifts of that pattern. See text for explanation. The longest ray is in the most favoured sector, activity in other sectors being shown proportionately. (After Sauer, 1957).

lings, which were direction-trained in a circular-choice apparatus with reference to the sun by day, took up the same direction when tested at night. In all these experiments with caged birds, good orientation was only observed on clear starry nights; overcast skies led to random scatter, and the level of activity fell sharply. Hamilton's ducklings were also confused during twilight, when the sun was below the horizon and the stars not yet bright.

The flight of wildfowl released at night can be followed by

attaching small electric lamps to their legs. Using this technique Bellrose (1958 b, 1963) demonstrated nocturnal orientation, in the same 'nonsense' direction as by day, in Mallard, Blue-winged Teal, Pintail and Canada Geese. Matthews (1963 c) in confirming this result with Mallard also showed that the same *individuals* could orientate under starry skies at night as well as by day. Both authors agreed that with overcast skies orientation was lacking; with partly clouded skies or high thin overcast the results were equivocal. Cochran *et al.* (1967) using the technique of bird-borne radio transmitters (p. 74) concluded that straight migratory flights could be made at night under overcast, but only if clear or partly clouded skies are available during the day or evening before departure.

There is thus a strong *prima facie* case for orientation directly by the stars. In one respect this could be simpler than orientaby the sun. At any one place, throughout the year, a given star rises on the same bearing, culminates at the same altitude, and sets on another fixed bearing. There is no seasonal rise and fall of the arc it describes across the sky. Variations through the night in the rate of change of azimuth, due to changing altitude between rising and setting are therefore the same the year round. But there are other complications, because of the difference between the length of the solar and stellar days. Both are based on one rotation of the Earth as measured by two successive transits of an astronomical datum across the observer's meridian. For the stellar day the reference point is on the star sphere, which for practical purposes is at infinity and unchanging in position relative to the Earth; successive transits occur at intervals of 23 hours 56 minutes. For the solar day the sun is the reference point and the Earth is moving in orbit round it in the *same* direction as its own rotation; successive transits therefore take 4 minutes longer, 24 hours (with minor fluctuations). The stars thus rise 4 minutes earlier each solar day (fig. 12), 2 hours earlier after a month has elapsed, and the star sphere thus appears to slip westwards until a year later it has regained its original position. Thus, at a given (solar) time, a bird would have to fly at a differing angle to a given star according to the calendar. However, the experimental evidence is that birds do not use a time-compensated azimuth orientation with reference to a star or cluster of stars. Matthews (1963 c) subjected free-flying Mallard to the same treatment that had

43

shifted the 'clocks' of birds from the same stock and induced changed orientations with reference to the sun (p. 32). But clock-shifted 6 hours forward, 6 hours back or 12 hours out of phase, and released under starry skies at night, they orientated

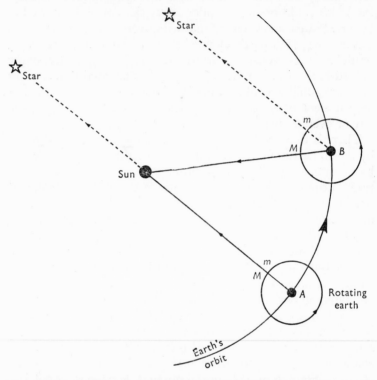

Fig. 12. Solar and stellar time. Solar days are measured by the successive transits of the sun past a given meridian (M) as the earth rotates. For stellar days the reference point is on the star 'sphere', virtually at infinity. Therefore as the earth moves round its orbit (*A* to *B*) a discrepancy develops due to parallax, the sun being relatively close. The stellar day is thus shorter (23 hours 56 minutes).

north-westwards just as strongly as did control birds (fig. 13, cf. fig. 7).

It is true that time can be deduced, at any one season, by measuring the radial movement of the constellations close to the axis marked by the Pole Star. But to locate that axis requires either prolonged observation to determine where movement is absent or interpretation of the alignment of the constellations, e.g. Ursa Major's pointers. But if Polaris can be found we have a perfectly good indication of north already. It seems likely that something similar to this method, which we use ourselves and which does not require a knowledge of time, is used by birds. Directional indications are by no means limited to the circumpolar area. For example, Orion's Belt is east when perpendicular, south when at 45° and west when parallel to the horizon; Cygnus on its side is NE, perpendicular it is NW; when the Great Square of Pegasus is horizontal, its centre is south, and it rises ENE and sets WNW; Castor & Pollux rise

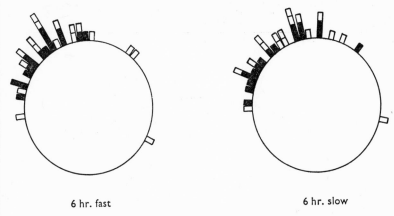

6 hr. fast                               6 hr. slow

Fig. 13. Time shifts of their internal 'clocks' do not change the 'nonsense' orientation of free-flying Mallard at night, showing it to be based on the pattern and alignment of the constellations. Compare fig. 7. (From Matthews, 1963c.)

NE and set NW; and so on. Some of the more prominent stars, such as Sirius, Procyon, Antares and Fomalhaut, also give good directional clues when near the horizon.

The westward slippage of the star sphere means that, although at any one place the same constellations pass across the sky in the course of 24 (solar) hours, different constellations will be visible during darkness at different seasons (fig. 14). Only the circumpolar constellations, which lie close enough to Polaris to pass above the horizon, are to be seen on every night of the year. However, Matthews (1963c) reported good orientation by Mallards under starry skies from September to May, through 300° of the star sphere's rotation. It is obviously going to be difficult to decide, on the basis of field experiments, whether

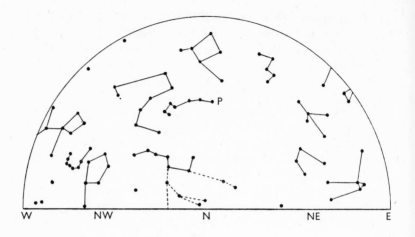

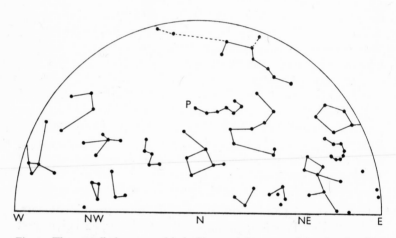

Fig. 14. The constellations seen when looking north in autumn (above) and spring (below) in north temperate latitudes. Note that the circumpolar (P = Polaris) stars remain visible.

the birds can use only the circumpolar constellations or a variety of constellations as they successively appear in the night sky. It is seldom that clouds cover certain constellations long enough for orientation in their absence to be demonstrated. Wallraff & Kleber (1967) have introduced an ingenious device which holds promise of isolating the orientating elements of the star pattern. A Mallard is strapped to a turntable and faced in various directions, in one of which it learns to expect a mild electric shock. Its reaction is detected by monitoring its heart beats. These increase in the learned direction under a starry sky but there is no reaction if a sheet of frosted plastic is interposed.

The use of a planetarium, the other possible approach, was pioneered by Sauer (1957). He demonstrated that Blackcaps, Garden Warblers and Lesser Whitethroats in Kramer-cages showed no orientation while the dome was illuminated with diffuse light, but when the star patterns appropriate to the place and time were projected the birds took up orientations essentially in their normal migratory directions. The star projector, in his model, did not rotate continually, so any interpretation of the constellations to ascertain the position of Polaris did *not* depend on observation of the rotation of the star sphere about it. Another limitation of his apparatus was its size, about six metres across; movements of the bird in its cage (90 cm diameter) thus induced considerable positional changes relative to the star pattern because of parallax, but apparently without upsetting the bird's reactions.

The Sauers furnished a series of positive reports on planetarium experiments (Sauer, 1961; Sauer & Sauer, 1960, 1962). A number of other workers are known to have failed to obtain consistent orientation results with migrants in the planetarium situation (e.g. Wallraff, 1966a). However, Emlen (1967b) was able to confirm that his Indigo Buntings did orientate under the artificial stars. He also carried out the critical test of reversing the projector so that the artificial Polaris was in the true south (instead of close to the north, as in the Sauers' experiments). The reversal of orientation which followed definitely confirmed that the projected points of light, not any outside influence, were of paramount importance.

By rotating the projector about its polar axis the stars appropriate for, say, 6 hours earlier can be provided. But the seasonal slippage of the natural star sphere means that these stars are also appropriate for the original time 3 *months* later. There is a third possibility that the stars are appropriate to the original time and season but at a location 90° east in *longitude*.

In the case of one Lesser Whitethroat (fig. 11) and one Blackcap subjected to skies one to six hours in advance, Sauer found what he called 'compensation' orientation. That is, the

47

birds turned to the west as if they were treating the sky change as due to their having been shifted abruptly in longitude, to the east. Wallraff (1960 *b*) detected an inconsistency in the White-throat results as reported, disorientation occurring at one stage instead of 'compensation'. Moreover 'compensation' east-wards did not occur when the birds were shown skies behind in time. The evidence is certainly not sufficient to establish Sauer's claim that the birds are capable of estimating their position in longitude from the aspect of the starry sky. Wallraff suggested that such deviations as were observed could be explained as shifts imposed on a simple time-compensated star-compass. This conflicted with the negative evidence of the time-shifted free-flying Mallard (p. 45). Furthermore Emlen (1967 *b*) found (in opposition to Sauer's results) that presenting his Indigo Buntings with planetarium skies shifted by ± 3 and 6 hours did *not* disturb their southward orientation.

When the artificial skies were more than 6 hours out of phase Sauer claimed his two birds, and another Blackcap, were dis-orientated (fig. 11). His interpretation was that the birds, in autumn, were being presented with skies appropriate to winter or summer under which migration does not normally occur. These skies were either confusing or lacking in orientation clues (with the implication, incidentally, that the circumpolar stars are not sufficient for orientation). When the sky presented was around 12 hours out of phase Sauer reported 'conflict' orienta-tion (fig. 11), the birds alternating between the direction appropriate to the current migratory season and the reverse direction of the opposite season. This was observed in one Black-cap in spring and in one Blackcap, one Garden Warbler and two Lesser Whitethroats in autumn. Sauer's interpretation was that the night sky of spring and autumn tends to stimulate the migration appropriate to those seasons as well as provide the directional information. Yet the birds would normally start spring migration under a very different star pattern, far to the south of the breeding latitude. Again, Emlen's results contra-dicted Sauer's, for he found no change in orientation when his birds were presented with star patterns 12 hours out of phase.

Comparing diagrams for 'disorientation' and 'conflict' one appreciates the difficulty of deciding where the boundary be-tween the two should be drawn. No statistical justification was offered by Sauer. Indeed the whole problem of the statistical interpretation of radial scatters is a thorny one, particularly the discrimination between comparative data. As Waterman (1963) has stated 'Unfortunately, powerful and well-developed methods for accomplishing this are not yet available to biologists'. Batschelet (1965) has provided a most valuable compendium of such methods as are available for dealing with circular distributions. The vector type diagrams used by Sauer are particularly likely to mislead in that his vectors are percentages of the time of migration activity in the most favoured sector. Generally the total time of activity was about 30 minutes, but, for example, a set of twelve, at first sight comparable, vector diagrams (Sauer, 1957, fig. 18) are found, on inspection of tabular data, to be based on activity periods ranging from 2 to 157 minutes. Kramer's diagrams (fig. 6) where each dot represents a given period of activity, make for a more ready comparison. But in either case it is important to remember that each diagram records the behaviour of only one bird in one experiment, *not* a series of *independent* choices by one bird nor the individual choices of a series of birds. It is very easy to get the impression that a lot more data have been presented in support of a hypothesis than is the case.

Some further claims made by Sauer on the basis of his planetarium experiments are better considered at a later stage (chapter 9). The main disappointment in the planetarium experiments is that little progress has been made in discovering which constellations are necessary for compass orientation. Emlen (1967b) reports preliminary experiments in blocking out portions of the artificial sky. These suggest that constellations within 35° of Polaris are important to the orientation process, whereas those in the southern half of the sky can be dispensed with.

In the long term the relation of star pattern to season is not so fixed as we have been assuming. This point has been considered by Agron (1963). The earth is not a regular sphere, but a slightly unbalanced spheroid. The gravitational forces of the sun, moon and planets, acting on the equatorial bulge, cause the spinning axis to describe a cone, though maintaining its inclination to the plane of the orbit round the sun (much as a spinning top will wobble if given a push from one side). Besides this steady precession of the axis, it also undergoes nutation, but the resultant sinusoidities in the axial movement are minor and of relatively short period.

We saw (p. 43) how the difference of 4 minutes in the length of solar and stellar day lead to the star sphere slipping westward relative to the solar day. Precession can likewise be thought of as being due to a difference of 50 seconds between the stellar year and the solar year causing the star sphere to slip westward relative to the solar year. Put another way, the Earth reaches a given inclination to the sun at a point slightly earlier on its orbit each year. After 13,000 years the stars that were visible (since they are opposite the dark side of the Earth) in spring will be seen in autumn (fig. 14). Only 6,500 years pass before summer stars are spring stars. Yet Sauer would have them to lack migratory stimulus or orientation guidance in the one case and innately to stimulate and direct the migration to the breeding grounds in the other. It is not impossible that genetic changes could be incorporated over such a short period in the evolutionary scale. It is more unlikely that the necessary changes could be made over and over again, returning to the original every 26,000 years. The whole cycle would have to have occurred some forty times since the Pleistocene (when most modern birds were extant). The stimulation to migration can probably be left to the rise and fall of the sun arc. The seasonal sequence, summer, autumn, winter, spring continues despite precession since it is produced by the movement of the earth relative to the sun alone. We could still accept that the stars could provide orientation information and that the same constellations are not necessarily used all the time. But in a long-term evolutionary trend the circumpolar stars would seem to have advantages. Here too, however, movement of the earth's poles relative to the star sphere causes changes, even in historic times. Thus Homer did not know Polaris as the Pole Star, it was then some 12° away from the axis of the rotation of the star sphere; nor was Cassiopeia then a circumpolar constellation to observers in the Mediterranean (fig. 15).

And what of the planets? While it is unlikely that they could be used for any detailed navigation because of the complexity of their movement relative to the earth, it is not impossible that they could serve as guides for (time-compensated) compass orientation. Venus is always and Jupiter often brighter than any star and could, for instance, provide directional information when the cloud blots out the stars. Venus is either on the south-west or south-east quarter according to the season. However, this is but speculation and we have no evidence one way or the other.

The moon, like the planets, could serve as a rough guide to position without detailed adjustments on the part of a bird. Moreover its phases furnish further clues, e.g. the first quarter is south early in the night, the full moon at midnight, the third quarter not until late in the night. As the moon is moving so close round the earth, changes in its relative position are correspondingly great—about 12° eastwards per solar day—and it

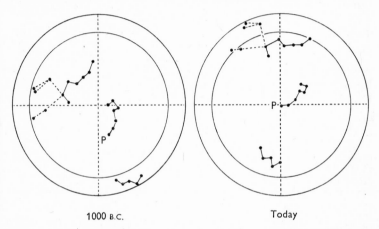

1000 B.C.          Today

Fig. 15. Variation in the positions of the circumpolar stars with time. The inner circle includes stars circumpolar at 40° N, the outer at 50° N. Homer would not have known W-shaped Cassiopeia as a circumpolar constellation, nor have had Polaris (P) to fix the position of the celestial Pole.

rises nearly an hour earlier each night. Time-compensated compass orientation is thus more complicated than in the case of the sun. Certainly the moon is not essential for bird migrants; the scale of their movements shows little correlation to the phase or visibility of the moon (L. Dinnendal, in Kramer 1952). Moreover all the workers with caged migrants under night skies (p. 41) were agreed that the presence of the moon if anything reduced the accuracy of orientation. It led to confusion or to simple phototactic responses to the side of the cage which was illuminated.

Hoffmann (1965) has recalled the 'boy scout' trick of telling the approximate direction of the sun from that of the moon. This requires a correction dependent on the phase of the moon

—when the moon is full the sun is opposite; at first quarter it is 90° right, at last quarter 90° left of the moon. On these bases it would not be necessary to postulate a separate lunar clock. However, Papi & Pardi (1959, 1963) provided solid evidence of time-compensated moon orientation in the amphipod *Talitrus* (whose eyes are unlikely to distinguish moon shapes, or the stars). On the other hand, Enright (1961), using another amphipod, *Orchestoidea*, concluded that time-compensated orientation only occurred on a night when the animals had observed sunset and moonrise. After being kept in darkness for several days they did no more than move at a fixed angle to the moon. This suggests an 'hour-glass' timing mechanism, set in motion at each sunset, rather than an oscillating lunar rhythm. Moon orientation of at least this degree of sophistication is indicated by the experiments of Ferguson *et al.* (1965) on the frog, *Acris*. Horridge (1966d) suggested that if rate and direction of change of altitude of the moon could be measured, compass orientation could be achieved without reference to any timing mechanism.

Working with free-flying Mallard from a population showing NW 'nonsense' orientation, Matthews (1963c) reported that clear night releases with a full to half moon gave rather poorer orientation than with stars alone. Possibly the moon, by hiding part of the star pattern in its glare, might actually make orientation more difficult. When Mallard with their clocks shifted 12 hours out of phase were released under stars and the moon either rising (last-quarter) or setting (first quarter), no deviation from the directions taken by controls could be seen. Thus if conflicting information was offered by the star pattern and by the moon position, orientation by the former apparently had precedence. Also these tests showed that their unwinding sun-angle mechanism did not lock-on to the moon as a sun-substitute; in the first case the moon was in the same, in the second case in a different, sector of the sky from the 'sun' position.

In the field the only critical tests of the value of the moon for orientation can be made when the stars are obliterated by cloud thin enough for the moon to shine through it. Suitable 'moon-only' conditions are somewhat infrequent (even in Britain), are hard to forecast, often accompanied by strong winds and generally of short duration. Progress is therefore slow, but sufficient releases have been done (Matthews, 1963c and unpublished) to show that good NW orientation *is* produced in moon-only conditions and with the moon azimuth varying from SE through to SW; this suggests strongly that fixed angle orientation is not enough.

Mallard clock-shifted 6 hours forward and released under moon-only conditions showed a (rather wide) scatter south-westwards. Now these birds had been in the artificial day/night regime, without sight of the moon, for seven days. In this period the moon would have slipped eastwards nearly a right angle. If this were not taken into account it would be in addition to the anticlockwise swing apparently induced by the time shift and the birds should have gone SE. That they did not do so suggests that they have sufficient of a nautical almanac 'in their heads' to tell them where the moon ought to be on any particular night at a particular time—not just roughly where it was at that time the previous night. Confirmatory experiments, with birds whose clocks had been 6 hours retarded and which were released when the moon should be above the horizon, are still not complete enough to be convincing. Some birds released when, according to their clocks, the moon should *not* have risen went off between SW and NW. Consideration of fig. 8 in terms of moonrise and moonset instead of sunrise and sunset, will show that such a result is explicable if there is a moon angle-correcting mechanism which *unwinds* after moonset. Moon orientation may thus be subserved by as complicated a mechanism as is sun orientation— a surprising conclusion in view of the relatively few occasions on which birds will have to rely exclusively on the former.

# CHAPTER 5

# Homing Experiments

When considering the experiments in which migrating birds were displaced to one side of their track (p. 17) we saw that the adult birds, instead of continuing in the original direction, tended to regain the normal wintering areas. These older birds, instead of passively accepting the experimental displacement, were in fact actively *homing* to a previously known area. This would pose very much greater navigational problems. The ways in which these are solved cannot be studied in detail by the displacement of passage migrants since, amongst other things, it will be necessary to know the precise area to which the bird is striving to return, and to maintain watch there to see if and when it returns. It is therefore necessary to force the birds to undertake long flights back to their home by removing them artificially and releasing them at a distance in areas unknown to them.

Such homing experiments have a respectable antiquity. Pigeons were certainly used to convey messages by the ancient Egyptians, the Greeks (Ovid) and Romans (Pliny) and generally throughout the Middle East. It was not until the nineteenth century, however, that the advent of railways enabled full use of the potentialities of Pigeons to be made. The practice of racing Pigeons for sport began in Belgium about 1825, and it is clear that the intensely selective breeding which has occurred over the intervening years has greatly improved the stock. The sport has spread to all parts of the world, and in Great Britain alone more than a million birds are maintained for this purpose. Many general accounts of pigeon-racing techniques have been given, including those of Tegetmeier (1871), Rivière (1923), Knieriem (1943), Nichol (1945) and Kramer & Seilkopf (1950), and there is a vast popular literature. Pigeons home only to the place where they were raised and not to that from which their stock originated, and 'home' does not appear to be permanent until they themselves have bred there. Up to that time they can be 'settled' in a new home, but thereafter this becomes extremely difficult.

For a time emphasis on one-direction flights (p. 17) resulted in a general belief that the learning and detection of such a direction was the limit of navigational ability in Pigeons, and that without it any homing could be attributed to a general search for known visual landmarks. The results of Heinroth & Heinroth (1941) and Platt & Dare (1945) supported this view. The work of others, Thauziés (1910), Rivière (1929), Gibault (1930) and Grundlach (1932) suggested that Pigeons were not entirely restricted to one-direction homing. There were fairly frequent reports of Pigeons homing from novel directions as a result of accident or emergency, as, for example, those listed by Osman (1950) and Kramer (1953 b). Matthews (1951 b) was able to show that pigeons of good stock could ignore directional training and give good homing from points at right angles or in the opposite direction. Kramer & St Paul (1952) also obtained good results with Pigeons that had been given a bare minimum of non-directional training. The feasibility of using Pigeons to study problems of bird navigation having been established, a plethora of experimentation with these birds followed, summarized by Wallraff (1967).

Now, the whole sport of pigeon racing is based on the recognition that there are wide differences in homing ability between Pigeons and that only a small proportion (5–10 %) are capable of achieving the really long distance returns at high speeds. Many researchers have tended to overlook this fundamental point and to treat any and all Pigeons as equal homing automata which have only to be released in sufficient numbers to constitute a test of a particular hypothesis. Matthews (1952 b, 1953 b) analysed his experimental sorties and was able to confirm consistent individual variations in homing ability. He also showed that one stock of Pigeons was markedly superior to another when used in the same test—even though they had been raised by fanciers living only 5 miles apart. Such differences in pigeon stocks were confirmed by Pratt (1955), Hoffmann (1959 a)—comparing German and English Pigeons —and Schmidt-Koenig (1963 e)—comparing German with American Pigeons.

The homing performance of a stock of Pigeons undoubtedly improves, up to a point, with the increasing number of releases they have experienced (Matthews, 1953b; Hoffmann, 1959a; Wallraff, 1959a; Schmidt-Koenig, 1963e). In part this is the result of the weeding out of poor homers, but the individuals also improve. As the number of releases appear to be more important than the total distance flown (Wallraff, 1959a), it is probably the ability to cope with the procedure of transport and release that is improved, rather than a knowledge of terrain.

Again, in pigeon races success is dependent on exploiting genetical superiority by training and racing the birds in the advantageous way, and a folklore of recipes has been built up. Mating and breeding increase the attraction to the loft, cocks and hens having equal merits as homers. In the 'widowhood' system the cock is deprived of his hen, then given a brief sight of her before being sent on a race, but prevented from copulating. That privilege is accorded him on his return.

It is clear that it is not essential for the Pigeons to be breeding, since young birds 3 to 4 months old home extremely well, while Oordt & Bols (1929) obtained good homing with castrated Pigeons. Clausen et al. (1958) compared homing performances at different stages of the reproductive cycle, over distances of 75 to 360 miles. They found that homing was improved when the birds were incubating or feeding young and that the best performances were obtained when the two stimuli situations were combined, birds still feeding young 2 to 3 weeks old and just starting to incubate a second clutch. Wallraff (1959b) and Schmidt-Koenig (1963a) did not find an improvement in homing after stimulating reproduction by artificially lengthening the photoperiod. The latter also injected estrogen, progesterone and prolactin without effect. However their tests were made at relatively short distances, under 45 miles. By the same token the finding that thyroxin injections or the plucking of flight feathers to simulate moult did not affect homing need not be given much weight in the face of a mass of practical experience that moulting pigeons do not perform well over long distances.

The association of food with home is probably important, as shown by Matthews (1952b). In the course of learning experiments (p. 87) Pigeons were introduced once a day into a novel situation where food was provided. Some took food on the first day, others not until the ninth. For thirty birds there was a strong correlation between this unwillingness to feed in strange surroundings and their homing ability previously demonstrated in field tests. It is possible to split the food/home association by training Pigeons to feed in a conspicuously marked basket placed at some distance from the loft. Two-way homing can then be achieved over short distances.

Birds other than Pigeons have seldom been used for message-carrying. Pliny tells of Caecina of Volterra who used painted Swallows to carry home the colours of the winning race horses. Desbouvrie (1889) claimed to have developed a strain of messenger Swallows but appeared to be unwilling to demonstrate them. In the Pacific, Frigate Birds were long used for inter-island communication by the natives. During the present century, however, there have been a large number of homing experiments, greatly varying in scale and value.

Many have taken place out of the breeding season, and particularly during the winter, when it is easy to trap large numbers of passerines. Indeed, many small experiments were prompted by a desire to get rid of birds that persistently retrap themselves to obtain the readily available bait. Matthews (1955 b) listed thirty-five species, mostly small passerines, released within 25 miles of the point of capture. To these may be added Bouchner & Sedivy (1959)—House and Tree Sparrows, Sargent (1959)— Eastern Tree Sparrows, Barthel & Creutz (1959)—Dunnocks, and Bub (1962)—House Sparrows. Return over such short distances would not seem to involve any special navigational problems and can be attributed to a simple search for known landmarks. Nevertheless, the proportion observed to return is generally low. The birds may have been migrating through the trapping area and so have no cause to return to it, and birds of the year might have as yet no fixed winter home. Even if the bird returns it is not necessarily attracted to the trap and the chances of it being observed again, let alone recaught, are accordingly reduced. These limitations also apply to those winter experiments in which twenty-four species were listed by Matthews has having been released in considerable numbers at greater distances, involving a real test of navigational ability. Very few returns were reported. This is not unexpected since birds were often required to fly northwards in the middle of winter, such as the Black-headed Gull returning from Zürich to Berlin (Rüppel & Schifferli, 1939—who also had successes with Coots). Petersen (1953) also used Black-headed Gulls, and Rüppell (1940) Goshawks. Substantial numbers of passerines were transported considerable distances by Hilprecht (1935), Creutz (1941–61), Schifferli (1936, 1943a) and Roadcap (1962). The overall results could be explained by chance factors. In many cases marked birds were still to be seen around the point of release weeks or months later.

Where game ducks are displaced from their winter quarters sufficient may be shot in the following season to indicate the proportion which regained the normal area. McIlhenny (1934, 1940) reported on such results with Mallard, Pintail and Green-winged Teal which in the United States have as main 'fly-ways' the Atlantic and Pacific coasts and the Mississippi valley. In this experiment 440 of these duck were taken from their winter quarters in Louisiana, in the Mississippi valley, and released on the coasts. There were ninety recoveries of which seventy-nine had returned to the Mississippi fly-way. No distinction was made between adult and juveniles or even between direct (same winter) and indirect (via the breeding grounds) recoveries. Bellrose (1958a) took these matters into account when shipping 895 drake Mallard from the Mississippi fly-way to the Pacific fly-way. Direct recoveries were mostly in the area of release (suggesting they were wintering when caught, not passage migrants (cf. p. 14)). Indirect recoveries, nineteen juveniles and twenty-one adults, showed that of the latter two-thirds returned to the original fly-way, whereas the former mostly maintained their westward displacement.

Mewaldt & Farner (1957) reported the indirect return of two Golden-crowned Sparrows to the wintering grounds after they had been accidentally released 700 miles away and 150 miles off their usual migration fly-way. Systematic winter displacements of this species, and of the related White-crowned Sparrow, from a back garden in San José, California have given some remarkable indirect returns (Mewaldt, 1963, 1964 and in preparation). Of 414 birds released 1800 miles ESE, twenty-six were recaptured the following season in the same ¼ acre garden (one, released in April, retrapped in June and again in October, probably returned direct). Of 735 birds released 2400 miles E, eight were recaptured the following season. Again adults performed better than immatures. The distances covered and the pinpointed returns are certainly striking and, as natural mortality is around 50 %, the actual success is twice that observed. Nevertheless the amount of *positive* evidence is small and difficult to raise to statistically significant levels, while the long time interval between release and return obscures any interpretation of factors that might have affected the returns.

The displacements imposed on young birds in the migration (p. 15) and wintering (above) phases of their first year of life seem, in general, to be carried through to their choice of nesting area. That is, like young Pigeons, they have no innate knowledge of the ancestral home and adopt as home the area in which they first breed themselves. This is confirmed by the experiments in which fledglings were reared in strange areas and by others where eggs are taken from one area and hatched and reared in another. In all cases any birds found breeding have done so at the foster home or at suitable areas to which they had wandered, not at the ancestral home. Species involved in addition to those mentioned in chapter 2 have been Shelduck (Schifferli, 1943 $b$), Common Gull (Schüz, 1938 $a$), Wood Duck (McCabe, 1947; Bellrose, 1958 $a$), Yellow-headed Blackbirds (McCabe & Hale, 1960) and Pied Flycatchers (Mauersberger, 1957). Löhrl (1959) found good returns to the foster home in Collared Flycatchers hand-reared to flying in aviaries and then displaced 55 miles south. In this case quite a short period of freedom, around two weeks before the migration, appeared sufficient to 'imprint' the locality on the birds. Here the same directional tendency would be appropriate to natural and foster home alike. Pinowski (1967) found that young Tree Sparrows, after transport over a very few kilometres, tended to return to their birthplace *when* they became mature.

All the really critical homing experiments have been done with actively breeding adults. Watch can then be concentrated on the nest and often the bird can be retrapped. The birds used for such experiments should be robust enough to withstand handling and a long, generally foodless, outward journey in a dark box. Incubation should be prolonged and shared by the sexes, so that the nest will remain functional and attractive during the homer's absence. A nest in a burrow is protected against predation and pilfering, and makes easier the catching and retrapping of the homing bird. If the birds nest in colonies, the collection of an adequate number is facilitated, and one watcher can keep a check on many nests. The Manx Shearwater is an excellent experimental animal for this purpose; nesting colonially in burrows, sexes sharing an incubation of

59

## TABLE 2. *Long distance homing experiments during the breeding season*

| Species | Numbers Used | Returned | Distances (miles) | Authors |
|---|---|---|---|---|
| Adelie Penguin | 225 | 43* | 118–1364 | Penney & Emlen (1964, 1967) |
| Leach's Petrel | 160 | 98 | 65–470 | Griffin (1940) |
| | 61 | 48 | 163–2980 | Billings (1968) |
| Storm Petrel | 10 | 2 | 125–340 | Lack & Lockley (1938) |
| Laysan Albatross | 18 | 14 | 1325–3995 | Kenyon & Rice (1958) |
| Manx Shearwater | 40 | 17 | 125–930 | Lack & Lockley (1938), Lockley (1942) |
| | 696 | 463 | 65–3050 | Matthews (1953c, 1964) |
| Gannet | 24 | 14 | 66–213 | Griffin & Hock (1949) |
| White Stork | 25 | 13 | 30–1410 | Wodzicki et al. (1938, 1939) |
| Lesser Black-backed Gull | 225 | 136 | 30–420 | Matthews (1952a) |
| Herring Gull | 13 | 12 | 47–276 | Goethe (1937) |
| | 164 | 152 | 62–872 | Griffin (1943) |
| | 88 | 34 | 30–315 | Matthews (1952a) |
| Common Tern | 80 | 36 | 94–456 | Griffin (1943) |
| Arctic Tern | 17 | 11 | 35–254 | Dircksen (1932) |
| Sooty Tern | 53 | 23 | 66–850 | Watson & Lashley (1915) |
| Noddy Tern | 66 | 33 | 45–850 | Watson & Lashley (1915) |
| Swift | 21 | 10 | 150 | Spaepen & Dachy (1952, 1953) |
| Alpine Swift | 38 | 12 | 1020 | Schifferli (1942, 1951) |
| Wryneck | 19 | 6 | 211–930 | Rüppell (1937) |
| Swallow | 56 | 21 | 242–1150 | Rüppell (1934–8) |
| | 86 | 70 | 28–409 | Wodzicki, Wojtusiak, Ferens (1934–7) |
| House Martin | 26 | 7 | 316–450 | Rüppell (1934–8) |
| Bank Swallow | 237 | 92 | 25–175 | Sargent (1962) |
| Cliff Swallow | 143 | 65 | 40–115 | Mayhew (1963) |
| Purple Martin | 10 | 10 | 32–234 | Southern (1959) |
| Bluethroat | 20 | 2 | 273 | Stimmelmeyer (1930) |
| Red-backed Shrike | 12 | 1 | 223–745 | Rüppell (1937) |
| Starling | 802 | 226 | 40–1150 | Rüppell (1934–41) |
| Red-winged Blackbird | 76 | 13 | 30–210 | Manwell (1941) |
| Cowbird | 226 | 39 | 41–305 | Manwell (1962) |

The distances shown include the maximum from which a return was recorded.
* Incomplete check for returns.

53 days, and capable of remaining on the nest for a fortnight, with an egg whose advanced embryo can withstand chilling for at least a week (Matthews, 1954). The homers are marked individually with numbered metal alloy leg-rings, and the only

unequivocal proof of return is when the bird has been caught and its number checked. Individual indentification is also possible by codes of coloured leg-rings and/or plumage marks. Watch is maintained, as continuously as possible, and Griffin (1952 *a*) has used, on a small scale, a radio-active 'watcher'. He attached a source of *gamma*-rays to a bird's leg-ring and placed in the nest a Geiger counter connected to a recording device.

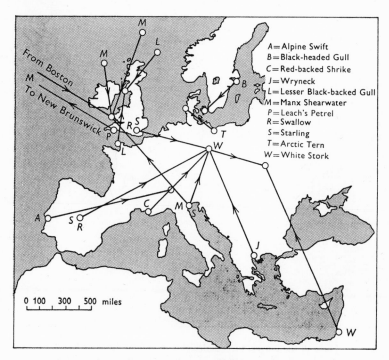

A = Alpine Swift
B = Black-headed Gull
C = Red-backed Shrike
J = Wryneck
L = Lesser Black-backed Gull
M = Manx Shearwater
P = Leach's Petrel
R = Swallow
S = Starling
T = Arctic Tern
W = White Stork

Fig. 16. Some of the longer, successful flights of homing birds in Europe. (Refs. Table 2 and p. 57.)

Matthews (1955 *b*) listed short-distance releases (under 25 miles) of breeding birds of thirty species. Again returns from such small distances do not imply any particular navigational ability. Nevertheless, they do provide controls for the long-distance tests, particularly as to the efficiency of the watching systems. To these may be added Leach's Petrels (Billings, 1968), Purple Martins (Southern, 1959), Mallard and Pintails (Heyland, 1965). The proportion of returns was high, giving confidence in the technique of checking employed. House and Tree Sparrows (Wojtusiak *et al.* 1947, 1953; Creutz 1949*b*) gave only 10 % returns probably because they were actually unable to return from more than very short distances. Table 2 lists the long-distance experiments with breeding birds, which provide the crucial tests of navigational ability. Only substantial

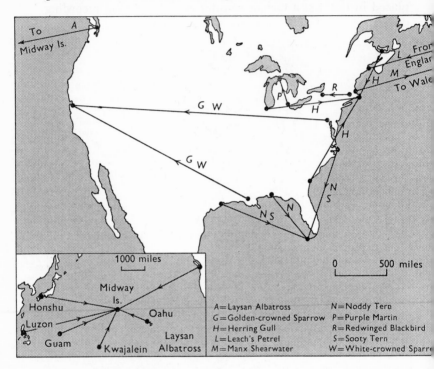

Fig. 17. Some of the longer, successful flights of homing birds in North America and (inset) the remarkable returns of Laysan Albatrosses to Midway Island. (Refs. Table 2 and p. 58.)

experiments giving at least one return from over 100 miles are included. Smaller scale tests of particular hypotheses or techniques are referred to later, some other trivial experiments were also listed by Matthews (1955*b*). The most valuable contributions are those in which large numbers of birds have been used, for only then can results be assessed with statistical confidence. Many remarkable individual returns have been achieved, some of which are illustrated in figs. 16 and 17.

Delving into the problems of bird navigation has its intellectual rewards, but there is an excitement in dealing with the navigators themselves that has little to do with scientific discipline. It was certainly not unremarkable to open a burrow on a Welsh island and find therein a Manx Shearwater ringed AX 6587, when that bird had been sent over 3000 miles to Boston, Mass. Added piquancy was given when the letter announcing its release $12\frac{1}{2}$ days before, came 10 hours later (Matthews, 1953c; Mazzeo, 1953). Reversing the process, one could not be indifferent to the fate of seven tiny Leach's Petrels when releasing them on the Sussex coast; nor fail to be relieved when a cable arrived reporting the return of the first two, in under a fortnight, to their nests on an island off Maine, again over 3000 miles away (Billings, 1968).

The longest successful homing flights to date have been those of Laysan Albatrosses taken from Midway Island in the Central Pacific. Two birds released in Washington State, a great circle distance of 3200 miles, returned in 10 and 12 days. Another actually returned from 4000 miles away, from the Phillipines, though taking 32 days to do so.

The sheer distances involved in some of these successful flights renders explanations of homing in terms of chance quite unrealistic. In addition evidence has accumulated following Matthews (1951b) and Kramer (1953a) that Pigeons, of good stock and in good weather conditions, will home from unknown points at straight-line speeds close to those of normal flight. This is especially so when the birds are in small groups. There may then be an element of competition, or, perhaps, less tendency to be diverted.

The situation is not quite so unequivocal in the experiments with wild birds. In the earlier tests, up to those of Griffin (1943), the birds returned at low speeds, on the average covering in a day only as many straight-line miles as they could have done in a few hours flying. There were isolated cases of birds returning much faster, even up to the same order as the normal flight speed. But, at the distances involved, these could have been chance exceptions. To cut down opportunities to rest on the way, Matthews (1953c) released the pelagic Manx Shearwaters inland. This species does not rely on thermal soaring, another possible time-waster en route. Of 152 released in sunny weather from mid-May to mid-June (conditions and timing shown to be conducive to good homing) 131 (86%) are known to have returned the same season, over half on the first two nights. This species does have the drawback that it returns to its burrow only after dark, to avoid predation by the larger gulls. Of the nineteen birds back on the first night (after releases 125 to 265 miles away) those released before noon gave homing speeds of between 9 and 20 m.p.h. (average 15·3 m.p.h.); those released in the afternoon clocked in at between 16 and 35 m.p.h. (average 23·9 m.p.h.). Clearly the other birds were arriving back off the island and having to wait before landing. Conversely, not all the later releases could get back in time to effect a landing that night. Thus from the 265 mile point twenty birds released in the morning gave six first night and two second night returns; twenty released in the evening gave no returns the first night but thirteen on the second. It seems probable therefore that many of the fifty-one birds arriving on the second night had homed very much faster than the extra 24 hours elapsed would suggest.

Similar arguments, that homing times include extensive periods of rest, and feeding, can be advanced to explain the relatively slow homing of other species. A critical, if rather unkind, experiment would be to release land birds over the ocean (Kramer, 1956). In this connection it should be remembered that breeding birds are not in their best physiological condition and certainly do not carry the stores of fat that enable them to make long sea crossings on migration. To get the highest homing speeds may require not only selection of the right time, place and conditions of release but also the best stock. There is evidence that, as in Pigeons, wild birds of different stocks have different homing abilities. Thus Griffin (1943) found that in the migratory Herring Gulls of New England more than three-quarters returned at speeds in excess of 60 miles per day; the near sedentary population in SW Scotland gave only 6% in that category (Matthews, 1952a). Mewaldt (1964) used two races of White-crowned Sparrows in his long-distance winter displacements. One, *pugetensis*, migrates north from California only to the Vancouver area and gave 7 out of 658 returns. The other, *gambellii*, migrates more extensively, to Alaska, and gave 14 out of 325 returns. Experiments with completely sedentary birds have been confined to House and Tree Sparrows, and these have given no returns from further than 9 miles (Wojtusiak *et al.* 1947; Creutz, 1949b). The homing Pigeon certainly derives from non-migratory wild stock, the Rock Pigeon, and is, of course, sedentary itself. However artificial selection for homing ability has been going on for centuries, probably replacing the selection which migration would impose. Gustav Kramer fell to his death while climbing to take Rock Pigeon eggs with a view to establishing a stock in captivity, to test just this point.

Wild birds, like Pigeons, will probably also show individual variations in homing ability within a local stock. By the time they are nesting adults ready to be used experimentally they will, of course, have been subjected to selection for such faculties during migration. Furthermore the repeated disturbance needed for a series of homing flights would lead to desertion if carried out in one season. A long-lived bird is therefore required and Matthews (1964) managed to get 144 individual Manx Shearwaters to make from two to four flights from different points. There were suggestive indications of consistency in performance. One bird in four successive seasons homed on the first or second night from 235 miles E, 265 miles NE, 115 miles NW and 160 miles W. The experiments did clearly show that homing performance improved with experience—of handling, transport and release in strange areas. The learning of landmarks at the release point or on the homeward journey was probably not involved, for improvement was just as marked whether the subsequent releases took place from the same point (or direction) or from others. Swifter homing could also be expected when the birds were taken at the beginning of their five day incubation shift. Their performance declined with the advancing season, those released in the second half of June onwards in sunny conditions gave only 65 % returns with a quarter back in the first two nights.

While we can thus, with some confidence, provide a recipe for swift homing flights that are convincing evidence for navigational ability, sight should not be lost of another possible explanation for the slow flights so often reported in the past. They could be due not to the birds dawdling and stopping along the straight line, but to their steadily flying much greater distances exploring for known landmarks. This idea appears to have originated from the observation that Pigeons circle the release point and Rennie (1835) suggested that 'they direct their flight in circles...a constantly increasing circle being made till they ascertain some known object enabling them to shape a direct course'. Heinroth & Heinroth (1941) claimed experimental support for this view, but their releases were at relatively short distances where, as we shall see later (p. 84), the more subtle forms of navigation are apparently inoperative. With a spiral search pattern average speed would decrease rapidly with distance but the proportion returning would not. This is contrary to results of those experiments giving rise to slow returns (fig. 18). Moreover, to fly a regular spiral over great distances would

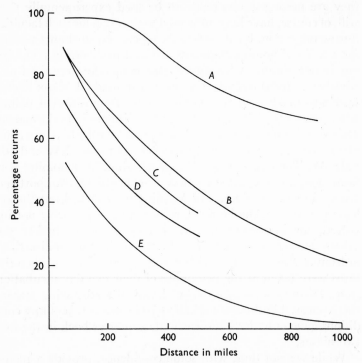

Fig. 18. The decline in percentage returns of homing birds with increasing distance. Various authors up to Griffin (1943). Herring Gull (*A*), Swallow (*B*), Common Tern (*C*), Leach's Petrel (*D*) and Starling (*E*). (Derived from Matthews, 1948.)

in itself require a high degree of navigational skill—an objection which applies to any type of systematic search or expanding search suggested by Griffin (1952*c*). Simple radial scatter in straight lines from the release point would only give a reasonable proportion of returns at short distances. The full potentialities of a completely random search for a known area were not appreciated until Wilkinson (1952) published an elegant mathematical analysis. He made the basic assumption that 'the mode of random search is to fly in a straight line for a certain distance, then turn, all angles of turn being equally probable, then fly another stretch in a straight line, turn again and so on'. He then fed in some perfectly reasonable figures for the parameters of flight speed, flight length, search time and target area and produced curves that had a startling similarity to those in fig. 18. His random search hypothesis was also able to reproduce the independence of average homing speed with increasing distance, or even its increase, which were common features of the earlier experimental results. Return speeds of the right order were also demonstrated.

Saila & Shappy (1963) demonstrated that random search combined with a small amount of directional (i.e. easterly) orientation could, theoretically, provide reasonable homing results in salmon returning from the high Pacific to their natal stream, given sufficient time. A numerical probability model (Monte Carlo method) and a random number generator in a high speed digital computer were used. Patten (1964) objected partly on the grounds of the highly restrictive assumptions required, partly because their model did not relate to biological processes. He proposed instead a model based on statistical decision theory which, with a low degree of rationality in the search process, was also sufficient to ensure high probabilities of return. Several workers, Adler & Adler (1966), and following R. L. Penney & G. S. Watson, Kendall & Speakman (1968) are investigating the Monte Carlo method in relation to bird homing, introducing into their calculations a relatively weak homeward bias. This tendency towards the home point is taken to be present at all times, not just at initial departure. Their conclusion, that the navigational mechanism need not be of very great accuracy to achieve substantially positive results, may well be of great relevance in understanding the survival value of incipient orientation tendencies. Kendall & Speakman predict that arrival directions depend on target size relative to a 'circle of confusion', where centripetal tendencies balance centrifugal tendencies inherent in random movement.

Wilkinson was careful to point out that the agreement between his theoretical predictions and the experimental results then available did not prove that random search *was* the

67

mechanism of bird navigation. In the light of the extremely long homing flights now reported, the evidence for straight line homing, and the homeward orientation to be discussed in the next chapter, it is even less likely that random wandering is even a large factor in homing. Nevertheless clinching arguments would be forthcoming if we could follow what a bird is doing in the gap between being lost to sight by its liberator and being caught back at its nest.

Matters would be made easier if birds did not fly. Hamilton & Hammond (1960) report unintentional experiments in which pinioned Canada Geese escaped from their pens in spring and were recaptured plodding northwards up to 25 miles away. Heyland (1965) wing-clipped a few nesting female ducks and found some walked back half a mile to the nest. He also followed, for short distances, flightless moulting adults and juveniles, without drawing any conclusions. The adults perhaps showed a trace of north-west 'nonsense' orientation. But these techniques are possible only in relatively unpopulated country and perhaps should not be encouraged. Penguins, on the other hand, make extensive migrations on foot and belly and Emlen & Penney (1964) and Penney & Emlen (1967) were imposing less by releasing them well inland in the featureless wastes of Antarctica. The birds, however, were not generally followed beyond a mile, and no track-marking device was used, although the observations of Sladen & Ostenso (1960) pointed out the possibilities. Radio-transmitters proved to have 'technical imperfections and irregularities' so they did not appreciably extend the information.

At least a partial answer would be obtained if we could determine how long the bird actually flew during its absence from home. Exner (1905) attached to Pigeons open-ended tubes containing camphor. The rate of evaporation of the camphor was increased by a flow of air and so, by weighing the tube on the bird's return, he hoped to measure the flight-time. But the evaporation was influenced by other factors such as temperature, and by the winds encountered on the ground and in the air, so no reliable results were obtained. Wilkinson (1950) devised a more refined form of flight recorder. This was essentially a small closed cylinder with at one end a source of radio-active *alpha*-particles, and at the other a sensitive photographic emulsion. The particles impinging on the emulsion make discrete tracks which could be developed. For a given source strength the number of tracks (counted under the high power of a microscope) was a direct measure of the time of exposure. Between source and emulsion was interposed a ball-shutter which only opened when the cylinder was horizontal. The cylinder was attached in such a way that it was horizontal when the wing was spread in flight, hence the time in flight would be measured. In practice (Matthews, 1953c) the device was found to have a number of snags, particularly the recording of spurious 'flight' time and non-uniform decrease in source strength, which greatly reduced its usefulness.

LeFebvre *et al.* (1967) constructed an apparatus for measuring the amount of time spent in the air by Laysan Albatrosses. Two electrodes protruding from the package conducted battery-produced electricity, which electroplated a third cathode, but only when the device was in the water. The weight of copper deposited was thus proportional to the resting period. A photoresister cell was also incorporated to register day length. The device has not actually been used on free-flying birds.

There is a little information to be derived from recoveries and sightings of homing birds en route. These must be treated with some reserve in case of bias by an uneven distribution of competent observers. Also birds recovered dead may have been weakened and were not necessarily still trying to home, though this objection need not apply to birds shot or crashing into overhead wires. Nineteen reports of wild birds found more than 25 miles from either release point or home, comprised eight Manx Shearwaters (Matthews, 1953c, and unpublished), two White Storks (Wodzicki *et al.* 1939), three Lesser Black-backed Gulls (Matthews, 1952a), two Herring Gulls (Griffin, 1943; Matthews, 1952a), one Sooty Tern (Watson & Lashley, 1915) and three Starlings (Rüppell, 1936, 1937). These recoveries and sightings were represented by Matthews (1955b, fig. 10), and support the conclusion of essentially directed flight. Two-thirds lie within 45° of the home bearing, and the average deviation for the whole scatter is only 47°. Ringing recoveries do not reveal how long the bird has been where it was found and so afford no indication of speed, but two Shearwaters and two Storks recovered on the day after release had covered respectively 129, 150, 142 and 180 miles in the direct line. Homing Pigeons are a special case in that lost or exhausted birds often enter a strange loft and could be reported through the head-

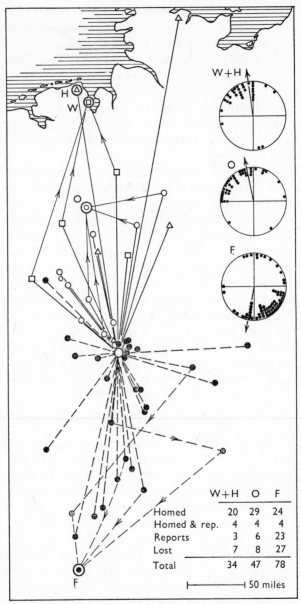

| | W+H | O | F |
|---|---|---|---|
| Homed | 20 | 29 | 24 |
| Homed & rep. | 4 | 4 | 4 |
| Reports | 3 | 6 | 23 |
| Lost | 7 | 8 | 27 |
| Total | 34 | 47 | 78 |

⊢————————⊣ 50 miles

Fig. 19. Recoveries of homing Pigeons from Hohenkirchen (H), Wilhelmshaven (W) and Osnabrück (O) to the north and from Freiburg (F) to the south of a central release point. The recoveries reflect the homeward orientations at departure shown in the scatter diagrams at the right. (From Kramer, 1959; after Pratt & Wallraff, 1958.)

quarters of the racing union. In England however there is little enthusiasm for this duty and Matthews learnt of the whereabouts of only a few of his lost birds. In Germany, fanciers appear to be more disciplined and Kramer and his co-workers regularly featured scatter diagrams of reports of lost birds. In conditions which gave good homing by the majority of the birds, even those falling by the wayside were preponderantly in the general direction of home. Fig. 19 illustrates one example in which Pigeons from four homes were released from the same point. In some later experimental releases with Pigeons having very limited flying experience (p. 135) few birds actually homed and the interpretation of the results was based almost wholly on the degree of scatter of the recovery points.

There have been a number of attempts to follow birds visually from slow-flying light aircraft. Griffin (1943) first used the technique on gulls and later on Gannets (Griffin & Hock, 1949). The tracking of nine Gannets for periods of up to 9 hours (fig. 20) was an outstanding achievement. The resulting tracks certainly suggest random or even spiral searching for landmarks, but Gannets make use of soaring on long flights and they may rather have been searching for suitable atmospheric conditions. Again homeward tendencies (and six of the birds started in the homeward direction) may have been baulked by high hills in the direct line of flight. Griffin (1964) has published a gripping account of the excitements of aerial tracking. Individual Pigeons have invariably proven to be too difficult targets to follow long, but several workers have had varying success with small flocks (Yeagley, 1951; Matthews, 1951 b; Griffin, 1952 b; Hitchcock, 1952, 1955). No very clear picture emerged from these attempts. Sometimes exploratory flights were apparent, at others more directed tracks resulted. Often topographical lines, of railways, roads or lake shores, diverted the birds much as leading-lines affect the standard directions of migrants (p. 3). But it was by no means certain that the Pigeons were unaffected by having an airplane closely following on their heels.

The use of radar to trace migratory movements naturally led to suggestions that homing birds should be tracked in this way. The response of the bird would not only have to be increased (by metallic reflectors) but would have to be made unique by means of a responder beacon on the bird. Radar signals are particularly likely to be swamped by ground responses when the bird is flying within a few hundred feet of the surface. Basically therefore, if a bird is to be encumbered with hardware it would better be in the form of a radio transmitter. This line has been taken up in America as technical advances in the production of extremely small components have made the project more feasible, though still expensive.

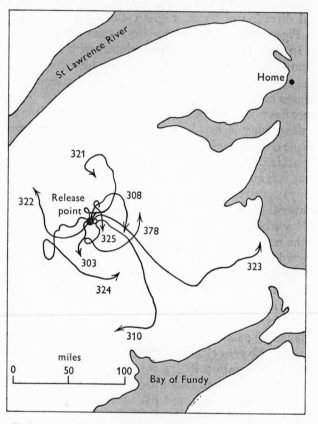

Fig. 20. Flight paths of homing Gannets followed visually from a light aircraft. (From Griffin & Hock, 1949.)

Lord *et al.* (1962) used a transmitter, weighing 38 gm, on a Mallard, but the main outcome was a measure of the respiratory rate. Southern (1964) was able to trace the movements of Bald Eagles in their feeding flights, sometimes over 38 miles in the day. In both these cases the receiving antennae were on the ground, with a consequent limitation in range, for the transmissions used could only be picked up when the bird was in line-of-sight, i.e. within 20 to 30 miles when flying at a few hundred feet. Even this limited amount of initial information would be better than nothing, but the tracking could be extended indefinitely if the receivers were mounted in an aircraft. This could then stand off, well away from the bird and so cause none of the disturbance inherent in close visual tracking.

Michener & Walcott (1967) have used such an airborne system, with a 28 gm transmitter harnessed to homing Pigeons which could be located by the direction-finding equipment with an accuracy of $\pm \frac{1}{4}$ mile. A total of ten birds were tracked over 131 trips. When birds were repeatedly released at a point 35 miles WNW from home in Massachusetts a striking feature was that no two tracks in 36 flights covered exactly the same route. The birds were therefore not 'steeplechasing' home, flying from landmark to landmark. Nor were they flying towards far distant landmarks since the tracks did not differ in character whether the visibility was 5 or 50 miles. There were some instances where Pigeons were apparently misled by a similarity between topographical features. This had been noted in the earlier investigation of Griffin and of Hitchcock (p. 72). However, landmarks *were* definitely used when close to the loft. A line could be drawn on the map round the home town beyond which its tallest buildings could not be seen, because of a ring of low hills, from the height at which the Pigeon normally flew (about 100 feet). Almost invariably when the birds reached this line they altered course to fly directly to the loft.

When Pigeons which had been repeatedly tracked from the west were released 30 miles north and 20 miles south they flew at first to the east. This showed that the previous flights had been by a form of learned compass orientation (p. 18). The trained direction was maintained for five or even occasionally fifty miles but then the birds would turn and fly directly towards the loft. This change was not associated with the birds coming within sight of previously traversed terrain and is thought by the authors to represent true navigation. This was seen many times and its accuracy was extraordinarily good, courses within $\pm 2°$ of the correct home direction being not unusual. Further tests on the bases of such navigation were carried out and are discussed in chapter 11.

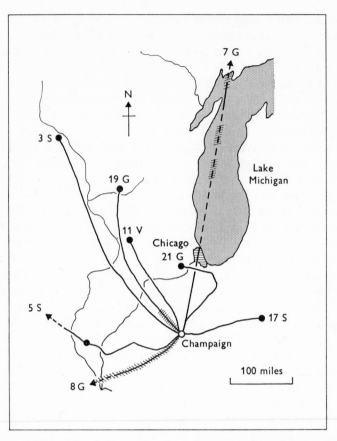

Fig. 21. Flight paths of nocturnal spring migrants carrying miniature radio-transmitters, from Champaign, Illinois. Flights under open skies except where paths are double shaded (⧘⧘⧘). A dot indicates that the bird landed, an arrow head that it was still flying when contact was lost. Major rivers, as well as Lake Michigan, are indicated. G = Grey-cheeked Thrush, S = Swainson's Thrush, V = Veery. Numbers are flight references. (After Cochran *et al.* 1967.)

Cochran *et al.* (1967) and their colleagues in Illinois have also been using the technique of aerial tracking of bird-borne radios (fig. 21). Indeed they have cut the transmitter down to a size where its weight is only 3 gm. This means that it can be carried by quite small passerines. One remarkable tracking was of a Swainson's Thrush that took off, in spring migration, at 20.00 hours and was shadowed through the night until it landed 8 hours later 450 miles NW. The actual track flown measured only 453 miles (3 S). Even if the bird was flying on a compass course, and not homing to its breeding quarters, such accuracy is the envy of the human navigator. In another case a Grey-cheeked Thrush was tracked for 400 miles. It took off at 19.55 under clear skies and flew NNE at 50 m.p.h. helped by a tail wind of 22 m.p.h. After 140 miles its course took it over Lake Michigan while the plane, for safety, had to go round by land. However contact was renewed in Wisconsin at 02.48, but then the plane had to turn back because of a thunderstorm (7 G).

Undoubtedly much more will be learned about the homeward flights of birds by means of sophisticated techniques such as these. It may not be so long before suggestions that bird-borne transmitters should be picked up by artificial satellites are adopted. Their position would then be fixed every $1\frac{1}{2}$ hours in the course of long migratory journeys. At present the batteries needed to give sufficient range (*ca.* 150 miles) make for a heavy apparatus, apparently suitable only for elephants—not the most convenient of experimental animals.

# CHAPTER 6

# Homing Orientation

Attempts have been made to quantify that other part of the homing performance most open to observation, the behaviour soon after release. If no indication of homeward orientation were forthcoming this would not exclude navigation (p. 117). But if definite homeward orientation could be repeatedly demonstrated in completely unknown areas, the case for navigation would be incontrovertible.

The study of initial orientation has, at least as a technique, the beauty of simplicity. Each bird is watched out of sight with powerful binoculars and the point at which it vanishes (and the time taken to reach it) noted. This may be misleading as an indication of flight direction for the individual. For instance, it may indicate one limit of a side-to-side movement, or an initial false start may have taken the bird so far that it is lost before its final heading approaches the same bearing from the observer. But with a large number of 'vanishing points' under a given set of conditions, a scatter diagram is built up whose orientation, or lack of it, can be accepted with some confidence. With two observers stationed a distance apart, cross-bearings can be obtained and so a representation of the bird's actual track is reached. Even with one observer, useful additional information is obtained by noting the bearing at fixed intervals after release.

The choice of release site is important; ideally it should be flat and featureless with an uninterrupted view in all directions. Release from the ground is preferable since it is easier to follow a bird silhouetted against the sky than when looking down on it from a high tower. To avoid imparting bias on release, the liberator throws the bird up into the air, facing successively in different directions.

Suggestions that homeward orientation occurred soon after release may be found in the work of Watson & Lashley (1915), Wodzicki et al. (1938), Schifferli (1942) and Wojtusiak (1949) but without sufficient numerical evidence. A first convincing demonstration of homeward orientation was obtained with racing Pigeons in 1949 (fig. 22). Thirty Pigeons released successively at three unknown points in radically different directions were consistently lost to sight predominantly in the

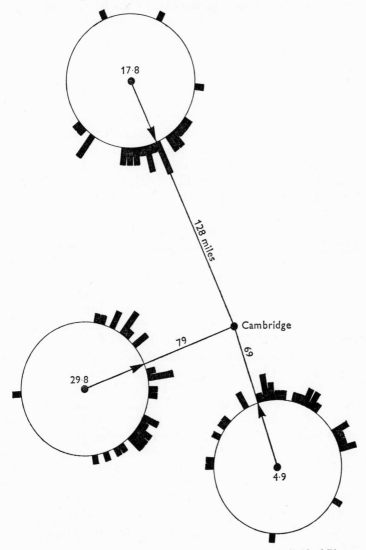

Fig. 22. Homeward orientation demonstrated by the same individual Pigeons in autumn 1949. Shortest spoke ≡ 1 bird. (After Matthews, 1951 *b*.)

homeward part of the sky (Matthews, 1951*b*). It was clearly important that such a crucial point should be established beyond doubt and Matthews (1953*a*, 1955*a*, 1963*b*) published over five hundred bearings from unknown points over 50 miles selected to take care of any possible bias due to training, topography, wind direction and the like. Of the bearings 56 % were within 45° of the home direction, as compared with 25 % expected in a random scatter. The accumulated data provided an impressive picture of homeward orientation (Matthews, 1955*b*, fig. 12).

Kramer (1952, 1953*a*, 1955) and Kramer & St Paul (1954), using Pigeons with little or no training, also obtained what at first seemed to be marked homeward orientation from a point 200 miles south of Wilhelmshaven. However, as work with the German birds proceeded it became clear that they had a strong directional bias; not only did they home much better from the south (Kramer, 1957; Wallraff, 1959*b*), but in their departure diagrams a distinct northward tendency was still apparent when the release points were west and east (e.g. fig. 23). Releases to the north were inconvenient, so further tests were made from Durham, North Carolina. Here Pratt & Thouless (1955) had already, albeit unwittingly, demonstrated a northward tendency in local racing Pigeons. Kramer *et al.* (1956) showed also that the Pigeons homed faster and more consistently from the south. The same authors (1958) found a similar directional bias in birds based at other localities in Virginia and North Carolina. Graue & Pratt (1959) found that Pigeons from a locality in California had a southerly bias while those in Iowa homed poorly from the west. Clearly orientation from the favoured direction in such cases cannot be accepted as evidence for homeward orientation. We would appear to be dealing in these cases with a form of 'nonsense' orientation such as has been discussed earlier (p. 18) in terns, wildfowl and penguins. Just as in those cases the orientation is near-immediate; the northwards orientation is obvious 20 seconds after release and the bearings at 40 seconds differ little from the final (fig. 23*a*, *b*). Such apparently swift homeward orientation was easily simulated by releasing Mallard from Slimbridge south-east of that place; the deception became obvious when the birds were released north-west (fig. 23*d*, *e*).

The Pigeons used by Matthews (above) did *not* show immediate orientation. Thus the birds shown in fig. 23c were, typically, still scattered round the sky 45 seconds after release, by the time Kramer's birds were nearly on their final bearing.

The time taken to reach the limit of visibility will of course be included in the 'time in sight'. This limit depends on a num-

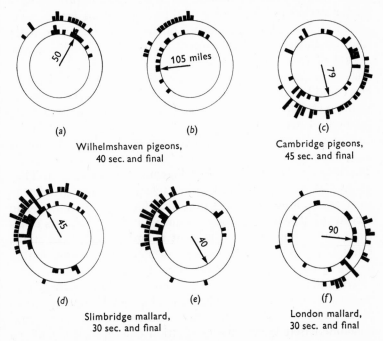

Fig. 23. 'Nonsense' orientation can simulate near-immediate homeward orientation if care is not taken to release birds from several directions. Thus Pigeons and Mallards in (a), (d) and (f) apparently headed almost at once towards home (inner circle, bearings at seconds indicated; outer circle, final bearings). Releases in other directions at (b) and (e) reveal a 'nonsense' rather than a homeward orientation. Note in (c) that these Pigeons showed no immediate orientation. (Derived from Kramer, 1957; Matthews, 1961, 1963a, b.)

of factors, eyesight and experience of the observer, optical aid used, background, clarity of atmosphere, direction of sunlight, colour of bird and its attitude to the observer—tail on or broadside on. But even when viewing conditions are favourable an experienced observer is unlikely to follow a Pigeon flying directly away from him for more than 1 minute. Matthews (1955b) showed that very few of his birds dallied for less than an extra minute. The birds lost quickly were quite well orientated and the accuracy of orientation did not increase steadily the longer the bird remained in sight. It follows that the birds dallying for an excessively long time were doing so for reasons unconnected with the orientation process. This does not necessarily mean that orientation would not have been improved by a longer period of observation if these extraneous factors were removed. Nor is it permissible to conclude that the birds quickly lost to sight were necessarily orientated by the end of the 'excess' period of time. We should expect a quarter to start out, by chance, within $45°$ of the home direction. The orientation process could be taking place while the bird maintained this chance trend, yet the appearance would be of orientation from the moment of release.

If swift homeward orientation were really a fact, we would expect to find individuals *consistently* making quick departures. The histories of 122 individual Pigeons were examined. They all had at least three fine weather releases at over 50 miles to their credit, amounting to 549 sorties. Of these birds 57 % had records of at least one disappearance within $2\frac{1}{2}$ minutes of release, but *not one* had a consistent record of such swift starts. Only ten had consistent records of starts within 3 minutes, i.e. of $1\frac{1}{2}$ to 2 minutes of excess time in sight.

To date consistent and *immediate* homeward orientation from various directions to the same point has not be demonstrated by any of the workers with Pigeons. Schmidt-Koenig (1965$b$) concluded that homeward orientation emerged after half the time in sight had elapsed (i.e. after nearly $1\frac{1}{2}$ minutes). The fact that some time is required before undeniable homeward orientation emerges is an important consideration when possible mechanisms of navigation are considered (p. 142). On the other hand, the existence of a simple compass type orientation immediately after release does not mean that true homeward orientation could not still emerge later. As Kramer himself recognized (1959) '...the observed northward trend might be the outcome of genuine homeward orientation plus non-specific northward tendency'. In a release from the favoured direction the compass-orientation and the homeward orientation would coincide and give the misleading impression of immediate homeward orientation. Releases from other directions should give evidence of a deviation from the 'nonsense' direction towards that of home at the extreme range of powerful binoculars and good eyesight. This is suggested in the diagrams of Pratt & Thouless (1955) from North Carolina and Matthews (1961) also mentions observing it there.

The experiment of Pratt (1956), in which Pigeons from Richmond were released west and south and others from Durham north and east, gave indications of final homeward orientation. There was also evidence of separation of vanishing points in the appropriate direction, at right angles, when birds from the two lofts were released alternately at the same point. These positive results were not achieved in other two-way releases in North Carolina, but releases in Germany of Pigeons from alternately north and south of the release point (Kramer *et al.* 1957; Pratt & Wallraff, 1958) gave a clear, homeward separation of vanishing points (fig. 19). Although the northerly birds were directed towards home by 20 seconds, the southerly birds were still scattered at random then, and their final orientation was considerably less sharp. Kramer (1959) reports another clear case of homeward orientation when birds were released 62 miles north-west and south-east of home. Again there is no indication of immediate orientation.

It should be pointed out that the homeward separation on release of birds from two lofts does not, in itself, rule out the possibility of confusion with 'nonsense' tendencies. Different stocks have different tendencies, even when the stocks were originally of common origin (it is by no means sure that the tendencies are wholly innate). A two-way homeward orientation was simulated by releasing Slimbridge Mallard (tendency NW) alternately with London Mallard (tending SE) at a point between the two places (fig. 23*f*). Apparent homeward separation was clear even by 30 seconds.

While there are different 'nonsense' tendencies in different stocks, it does not follow that all stocks and all species *must* have some directional bias or other. Thus Sargent (1962) found none in Bank Swallows and Matthews (1964) none in Manx Shearwaters. It is therefore quite probable that the stock of Pigeons used at Cambridge, England by Matthews (p. 78) was indeed free of these tiresome tendencies. There was also a marked difference between the way in which his Pigeons were prepared for crucial experiments, compared with those in Germany and North Carolina. The latter were given at most a minimal training by releases a few miles in various directions from the loft. Matthews adopted the traditional method of experiencing his birds by a series of releases at increasing distances, up to 80 miles, in one direction, generally north. This has the advantage of giving extensive experience in distance homing though the birds cover only a strip of territory. Time consuming and tedious though this may be, often resulting in losses, one is finally left with the real navigators for testing. There is the risk that the training direction would be so ingrained that the birds would fly only in that direction (p. 18). In an early experiment with a different stock of inferior Pigeons (from Norwich) Matthews (1951 b) did obtain such results; so did Kramer & St Paul (1950), though here with training from the south they could have simply been reinforcing the since-revealed 'nonsense' tendency. Graue (1965 a) in a carefully planned series of releases from twelve points around his loft (at only 10 miles distant) showed that a single release can cause quite a large deviation in that direction in the next release from another direction. The fact that the Cambridge birds did *not* show such an imposed direction *might* mean that by good luck a balance between opposing tendencies had been achieved. However, completely untrained Cambridge birds did not show any immediate orientation (Matthews, 1953 b and unpublished), while Hoffmann (1959 a), working at Cambridge, demonstrated no definite directional tendencies in English birds of similar origin. It is, however, unlikely that he would have done so, relying solely on differences in the time to return from 7 and 14 miles and not recording flight directions on release.

The distance of the release point from home has very considerable relevance to the study of homeward orientation. Matthews (1955 b, 1963 b) showed that a team of thirty highly experienced Pigeons, which gave good homeward orientation at 50/100 miles, failed to do so at various points 25/35 miles away. Yet the same birds showed good orientation from $2\frac{1}{2}$ to 18 miles from home. His interpretation was that the mechanism of distance orientation only operates down to somewhere around 50 miles, then there was a zone of country in which the birds had no means of direct homeward orientation, and finally, inside that zone they could orientate from remembered landmarks. Schmidt-Koenig (1963 c) also reported a tendency for better orientation with increasing distances and then (1966) fully and extensively confirmed the concept of a zone of disorientation. He carried out, in North Carolina, sixty-seven releases, each of twenty different and experienced homers, at release points in all four cardinal directions in eight distance groups, from 2 to 250 miles (except east in the latter case). The distribution of vanishing points of each release was processed to give a mean

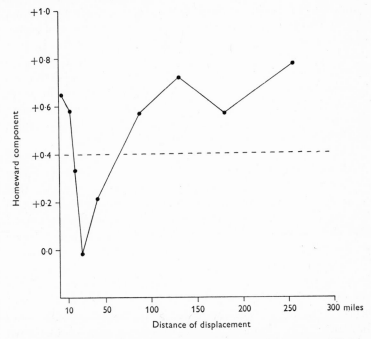

Fig. 24. The effect of distance of displacement on homeward orientation of Pigeons. Each point except the last (three releases) is based on eight releases of about 20 birds a time, in the four main compass directions. Homeward orientation is only apparent very close to home and above about 60 miles. In between it collapses. ($h > 0.4$ for significance.) See text for explanation. (After Schmidt-Koenig, 1966.)

vector (of length $a$ and direction $\alpha$) of the homeward component $h$ ($= a \cos (\alpha - \beta)$ where $\beta$ is the home direction). For the releases at each distance group $h$ was averaged arithmetically (fig. 24). Homeward directed orientation was indicated below about 12 miles and above 60 miles. In between there was a zone in which the homeward component is insignificant (headings widely-scattered or ill-directed and inconsistent).

The existence of a zone of disorientation limits the types of mechanism which could be responsible for homeward orientation at a distance (p. 94). For the moment there are some practical considerations. It is clear that releases of Pigeons less than 50 miles from home are going to tell us little about the processes involved in true navigation and must be discarded for this purpose. Indeed Mittelstaedt (in Schmidt-Koenig, 1963e) calculated the homeward component ($h$) for many hundreds of experiments with naïve Pigeons made by Kramer and by Wallraff in Germany and found little evidence of homeward orientation ($h = 0\cdot1, 0\cdot2$) until about 60 miles. In particular the experiments, involving thousands of Pigeons, in Germany to investigate the effect of various external factors on homing behaviour must be considered irrelevant to the main problem, since they largely concerned releases at these shorter distances (Kramer, 1954; Kramer & St Paul, 1956a, 1956b; Hoffmann, 1959a; Wallraff, 1959a, 1959b, 1960a).

The strategy of concentrating on short distance releases (with their saving in time, transport costs and losses of birds) probably stemmed from the conviction (Kramer, 1961) that '...it can be assumed that the orienting mechanism used in homing over short distances is used over long distances as well' and that landmarks played no part in either. This was based on the observation that the same type of orientation behaviour, interfered with by the same factors, was observable close to home as well as at a distance. The snag, as is now clear, is that the orientation behaviour observed was probably of the 'nonsense' compass orientation type. Thus Kramer (1957) cites homeward orientation at $7\frac{1}{2}$ miles from home which broke down with overcast skies. But the orientation was in the typical, northward 'nonsense' direction. Such orientation has been demonstrated in Mallard only one mile from home and likewise been shown to break down with overcast (Matthews, 1962 and in preparation). The existence of 'nonsense' orientation at short distances would also result in differences in speeds of return from short as well as from longer distances (Kramer, Pratt & St Paul, 1956). 'Nonsense' orientation at short distances based on the sun-compass would also be switched by time-shifting experiments, as indeed has been shown with Mallard at one mile from home (Matthews, in preparation). Even if time-shifted Pigeons released at a similar distance (Schmidt-Koenig, 1965a) had been originally home-directed this would still not preclude the involvement of landmarks. There could be map-and-compass navigation in the usual sense (not in Kramer's (1953a) looser sense). The bird could decide from known landmarks that it is, say, west of the loft and then fly east by its sun-compass. That this is indeed a likely process has been shown by Graue (1963). He released Pigeons between half and one mile from a loft in Ohio, N, E, SE and WNW. Control birds headed directly towards the loft. So did birds time-shifted 6 hours and released at points (E and SE) from which the loft was *directly* visible. From the N and WNW points the loft was hidden by woods and the time-shifted birds tended to switch left or right of the home line according to whether their clocks were fast or slow. Those birds shifted forward and released north, and so flying east, soon came into view of the loft and then turned abruptly towards it. One would like a repeat of this experiment in another situation where the shielding of the loft was reversed, to make sure than the birds were not adopting a broadly south-easterly 'nonsense' orientation out of sight of the loft and it was this that was being switched. On the face of it, though, Graue's interpretation, that both landmarks and sun-compass were involved, appears plausible, particularly when we remember the extremely short distance involved. It would help to explain the existence of a highly-developed sun-compass in non-migratory birds (and in those which migrate at night) if such 'translation' of landmark observation into compass directions was part and parcel of everyday life.

Homeward orientation at short distances in tortoises (*Testudo*) was affected by overcast and mirrors (Gould, 1957). This can no longer be considered to indicate that the same navigational process is at work at all distances. More likely, phototaxis was involved (Gould, 1960) or a sun-compass mechanism disturbed.

Graue's technique offers a means for confirming how far from home landmarks are recognized. If there are no innate or directional tendencies, time-shifted birds should be disorientated when the intermediate zone is reached. It is obviously more difficult to manipulate landmarks for Pigeons than, for instance, it has been in studies with bees or hunting wasps. Whitney (1963) reports an interesting experiment at a loft in wooded country. Young untrained Pigeons had been loath to fly any distance from the loft; trained birds had suffered heavy losses apparently through not knowing when to break away from the stream of racing birds passing along the coast 6 miles away. A 95 foot galvanized steel tower, surmounted by a large golden ball, was then erected near the loft. Thereafter young birds ranged widely (thus becoming familiar with more territory) and racing losses were sharply reduced. There are marked differences in homing performances of consanguineous Pigeons subject to the same training and released at the same place but homing to lofts a few miles from each other (Schmidt-Koenig, 1963c). These may well be due to differences in the ease with which the lofts can be sighted. Once Pigeons have been 'settled', that is, have adopted one locality as home, it is notoriously difficult to get them to accept another.

The ability to learn landmarks can also be tested by repeated releases in the zone of disorientation. This was done by Matthews (1963b) using, generally in sunny conditions, eighteen experienced Pigeons 25 miles north of home. On their second release 3 months after the first, a rather poor but definite homeward scatter resulted. The third release was with heavy overcast and gave a near-random scatter; the fourth, 3 months later, gave orientation as good as usually obtained at 50 miles or more; the fifth was very well orientated as was the sixth and also the seventh, this last again with heavy overcast. We can say that by the fourth release orientation by landmarks was established, as it might have been by the third—if the sun-compass had not been put out of action. By the seventh the knowledge of local landmarks was so good that the birds were able to, as it were, steeple-chase their way home. They acted as if they recognized that home was to be reached by flying towards landmark A, then towards landmark B and so on—rather than, as would Graue's Pigeons, that the release locality was north and home to be reached by flying south, with reference to the sun. That a southward tendency had not been trained into the birds was checked when they were next released, with sun, 23 miles west of home; they scattered at random. Another test that really well known landmarks can be used independently of the sun-compass was when the Pigeons were released $2\frac{1}{2}$ miles from the loft under thick overcast. They showed very marked homeward orientation.

While it does appear to be established that the learning and recognition of landmarks around the loft play a part in the terminal process of homing (p. 73), there are many indications that they have little to do with homeward orientation at a distance, nor with the homeward flight over unfamiliar territory (see also p. 65). Matthews (1952 *b*) checked this by taking twenty-four Pigeons with considerable homing experience (123 flights from 50 miles and over between them) and training them to associate food with a white card placed in front of one of a circle of eight covered pots. Learning was slow, from nineteen to a hundred trials being needed to reach a standard of eight successive correct choices. No correlation was found between the rapidity of learning and the excellence of the birds' previous orientation and homing performances. The birds were tested on the same problem 1 and 2 years later, giving a retention, measured as percentage reductions in trials taken to reach the same standard, ranging from 3 to 92 %. Again there was no correlation between excellence of memory and previous field performance.

Indeed it is clear that homing orientation is well developed without any experience of distant releases. Matthews (1953 *b* and unpublished) found that young Pigeons, hitherto having only flown in the neighbourhood of the loft, showed a marked homeward orientation when released singly at three points, SSW, N and SSE, 50, 60 and 78 miles from home. Fifty-four such birds gave 52 % bearings within 45° of the home line, a perfomance little inferior to experienced Pigeons. Pratt (1955) also reported homeward orientation in untrained Pigeons, but trained birds were released alternately with the untrained. Wallraff (1967) published a summary diagram of 578 departure bearings of Pigeons which had never been displaced by man before. The birds were released in twelve different places 53 to 102 miles from five different home sites. The number of birds from opposite directions was always equal. The general homeward tendency is strong (P $\ll$ 0·0001), though deviations of the vectors for individual releases are apparent (Wallraff, 1968).

Kramer & St Paul (1954) and Kramer (1957) reduced the experience of Pigeons still further by keeping them in a large aviary, 6 × 10 × 3 metres high, up to the time of release. Of eighty-eight Pigeons released 93 miles south, thirty-seven showed decisive departures and gave a clear overall tendency towards home (86 % ± 90°). Again we have the probability of confusion with the northwards tendency of the Wilhelmshaven stock. However, twenty-five Pigeons reared in an aviary at Osnabrück

87

and split between simultaneous releases 62 miles NE and SW orientated definitely homeward in both cases. None of these birds homed, and only four of Matthews', but fifteen of the Wilhelms-haven birds did so and a short distance (14 miles) release of such aviary birds (Kramer & St Paul, 1956$b$) gave fifteen out of twenty-one homing.

There would thus seem to be a dichotomy in the homing process, between the ability to orientate homeward and the ability to complete the homeward journey. The former shows little improvement with experience of flights away from the loft though in the first few flights a reduction of scatter may be found as well as some shift of bearings towards the home direction (Schmidt-Koenig, 1965$a$). The latter on the contrary is greatly improved (p. 56). Matthews (1953$b$), analysing orientation and return performances of Pigeons, found a wide range of individual indices and a difference in their distributions. That for orientation indices approaches the form of a normal curve, which would result if we were dealing with a single factor (or group of interdependent factors) ranging about the mean. On the other hand, the distribution of the return indices departs strongly from normal and is of the form that would be expected if we were measuring the effects of a number of independent factors. It was further shown that birds with poor orientation ability produced poor returns. With moderate or good orientation ability the other factors governing the return assumed the greater importance. This is in accord with the view that the orientation mechanism is not of phenomenal accuracy, but sufficient to impart a homeward trend to the flight, home being pinpointed by pilotage rather than navigation.

Kramer (1957, 1959) further restricted the experience of aviary-raised Pigeons by depriving them of any view of the surrounding landmarks. This was done either by placing the aviary in a large bomb crater conveniently left by the R.A.F. or by erecting an opaque palisade all round and slightly higher than the aviary. The results from the two experimental arrangements were very similar. Of 133 released at the south points, fifty-three flew well and still gave a homeward (or northward 'nonsense') orientation though less marked than in the open aviaries (70 % ± 90°). But the sixty-eight subsequent recoveries gave no indication of homeward trend and not a single bird homed. The fact that the initial orientation was present but not followed subsequently by homeward scatter of recovery points certainly suggests that in this stock the initial orientation is a 'nonsense' one (cf. p. 79). The exclusion of horizon and landmarks seems to cause a complete breakdown of homing ability. However, it is unlikely that the effect is due to the 'target' of known landmarks being reduced to nil; if this were the case, actual homings might be nil but recoveries would still be expected in the general area. Nor had the birds lost the ability to fly any distance since many were recovered over 50 miles from the release point. To examine the idea that the birds were mentally 'crippled' Kramer tried arrangements whereby different sectors of the horizon were visible to the birds. These experiments, completed by Walraff (1966 b), are more suitably discussed later (p. 135).

Returning now to consideration of homing in wild birds, definite homeward orientation in sunny weather was demonstrated in Lesser Black-backed Gulls (55 % of forty-eight within 45°) by Matthews (1952 a). Gulls were not very suitable subjects, a large proportion landed soon after release and they were influenced by soaring conditions to some extent.

Manx Shearwaters were then released by Matthews (1953 c, 1955 b) from a number of points, all but one inland and hence unknown to this exclusively pelagic bird. The points were arranged in all directions from the colony, except to the south where the coast would be well known to the birds, in case there was any directional bias by these birds when released inland. The exceptional point was on a weather ship in the open Atlantic, without any landmarks at all. In fig. 25 the vanishing bearings (to the nearest of thirty-two compass points) of 207 birds are grouped according to the three main directions of release points. The homeward trend is obvious (62 % within ± two compass points). Matthews (1964) has also shown that thirty-seven individual Shearwaters released a second time from a different direction shifted their departures broadly according to the new home direction. Previous experience of a homing release did not lead to an improvement in orientation. On the other hand, homing performance considerably improved (p. 65) and we appear to have the same dichotomy in the homing process as in Pigeons. This is also indicated by the experiment of Rüppell & Schein (1941), who kept young Starlings in aviaries until they nested in the spring. Released 72 miles away, none returned.

Other Starlings which had made a return migration before being so caged, homed quite successfully (orientation data were not available).

Sargent (1962), using Bank Swallows in Wisconsin, found strong evidence of homeward orientation at release points 1 to 25 miles from the colony. At points over 25 miles no significant orientation resulted and certainly at 50 to 100 miles the vanishing points were wholly random with regards home but with a strong tendency to go down wind. There are two possibilities that could

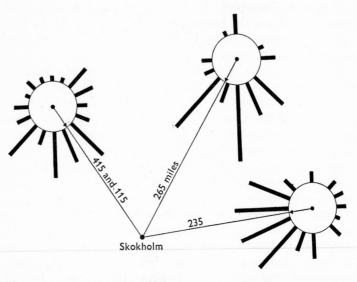

Fig. 25. Homeward orientation of Manx Shearwaters released in sunny conditions, vanishing bearings gathered into compass points. Shortest spoke ≡ 1 bird. (After Matthews, 1953c, 1955a.)

only be resolved by releases at greater distances; either the Bank Swallows have a wider zone of disorientation than racing Pigeons or they completely lack any navigational ability and depend wholly on landmarks for homing. As they are pronounced migrants this latter seems unlikely. The dependence of the short distance orientation on landmarks was shown by confining birds in a circular cage on a hill or roof top. The sector in which the bird was located was recorded at five-second intervals for the first ten minutes. Homeward tendencies were observed up to 25 miles and were absent thereafter. When the sides of the cage were opaquely screened the birds' movements were at random even at sites 1 to 10 miles from home.

The technique of studying orientation in caged, displaced breeding birds has been little used, largely because of the logistic difficulties of transporting the apparatus and trained observer, as well as the birds, to distant sites in various directions. Precht (1956) reversed the process and brought Black-headed Gulls from colonies in various directions to a closed room. Here each bird was placed in a circular wire-netting cage and the direction of its escape-attempts noted. Precht claimed that of 554 attempts by nine birds 95% were within the homeward half of the cage and 62% within the sector ±30° of the home direction. His pupil Gerdes (1962) produced results which, on one statistical assessment, differ little from chance— of 11,838 escape-attempts by 446 adult gulls, only 58% were in the homeward half of the cage. What is odd, however, is that there were concentrations in the sector ±30° about the home direction (40%) *or* in the opposite sector, ±150° (27%). Thus the escape-attempts were not distributed at random, but this by no means proves that there was any homeward orientation. We should not be over-impressed by the total of escape-attempts. The initial attempt is the critical one, subsequent attempts may well be influenced by what has gone before. The sequence of attempts was not given. In view of what was said earlier (p. 49) about the difficulties of statistically analysing circular distribution, the unlikelihood of reaching decisions as to preferred directions in apparently bipolar distributions will be apparent. Gerdes found the same type of distribution of escape-attempts whether the cage was in a closed, non-homogeneous room or in a hexagonal tent permitting only a view of the sky immediately overhead; and whether that sky was clear or overcast. However, if the lighting in the tent was at all unbalanced the birds showed obvious phototactic behaviour; if the test point was close to an arm of the sea this clearly affected the orientation when there was a breeze or when the bird had been at the test point for some time. Again all save two of the test points were within 40 miles of home in the zone where Pigeons, at least, would be disorientated or where homeward orientation would be by landmarks (p. 82).

Perdeck (1967*a*) used an apparatus that automatically

registered the position of the birds through pressure on 16 treadles round the cage. He found no homeward orientation in Black-headed Gulls taken from four breeding colonies when tested in a uniformly lit room or under a heavily overcast sky. When sun *and* landmarks were visible a marked orientation to the colony 15 miles away resulted. With regard the other three colonies 43 to 93 miles away, the headings were widely scattered though not wholly random.

Sauer (1963) had previously used a similar apparatus to get away from the possibilites of observational error or bias. In Alaska he hand-reared six Golden Plovers and then shipped them via Wisconsin, eventually to San Francisco. Their orientation was NNW, appropriate to the great circle course from San Francisco to the Alaskan home site. But a simple compass course from the normal wintering area in the Hawaiian Islands would be north, so that minor variation in orientation is not over-whelming proof that these birds (which had had no migration, or even free-flying experience) *were* showing homeward orienta-tion. Much greater separation of the two theoretical directions is needed. Also there is the difficulty that an apparent demon-stration of homeward orientation in naive immature birds runs counter to the massive field experiments (Chap. 2) in which only limited one-direction navigation is apparant. In this con-nection we might further note that Sauer claims that some of his birds were showing in their second autumn a south-east rather than a southerly orientation, as if they had adopted Madison or San Francisco as their winter home. However this was when they were exposed to the San Francisco skies although on an Alaskan time schedule—which should have been con-fusing to the birds if not to us. Rather extreme caution is needed in interpreting such results obtained in activity cages, especially when the birds were shown to be deflected by wind, passing clouds or aeroplanes. Otherwise an orientation can be produced for every combination of circumstances and any orientation can be explained one way or another. (See Arnould-Taylor & Malewski (1955), for instance.)

Recently, Dolnik & Shumakov (1967) subjected Kramer-caged migrants to large displacements from the Baltic (21° E). Under day and night skies, Barred Warblers changed their SE autumn orientation to S at 69° E and to SW at 135° E. These headings coincided with the E. African winter quarters, as if correction were made for displacement. Juveniles, as well as adults, were said to show the re-orientation. However similar changes would result from a sun-compass lagging behind the new local times. (This would not apply to constellation-orientation). Scarlet Grosbeaks, also night migrants, gave equivocal results, as did day-migrants, Chaffinchs and Starlings. Evans (1968) tested, in Kramer-cages, Scandinavian juvenile migrants drifted by easterly winds to England in autumn. Under clear night skies a Whinchat, 2 Redstarts, a Blackcap, 3 of 5 Garden Warblers and 2 of 3 Pied Flycatchers, preferred directions (SE to SSE) easterly from their standard directions (SSE to SW), as if off-setting displacement. Evans postulates that night-migrating juveniles are less likely to be influenced by passing conspecifics, when displaced, than are those proceeding by day.

Even if juveniles can correct for displacement, it does not follow that they have innate information on the *location* of their winter quarters. 'Compensation' orientation could be towards the place from which they were displaced. This orientation, vectorized with the standard direction, would produce a heading towards the still unknown goal.

In conclusion we may repeat that homeward orientation has definitely been demonstrated at considerable distances from home and independently of learned landmarks. This, together with the evidence on homeward flights, speaks strongly for the existence of a true navigational ability. Directional biases, innate or learned, can obscure the true homeward orientation and it is doubtful if it is shown closer to home than about 50 miles. At much shorter distances (up to about 20 miles depending on individual experience) homeward orientation does appear to be based on the recognition of learned landmarks. It is possible that the direction then taken up is determined through the operation of the sun-compass.

# CHAPTER 7

# Theories of sensory contact with home
# and of inertial navigation

Now that we are in a position to assert the existence of a true navigational ability in birds, as opposed to a simple compass-like orientation, we may examine the theories and evidence as to its nature. Random search may play some part in the return of birds released in the zone of disorientation (p. 83), but in general we must look elsewhere for the solution of the problem of homing. In our search we may well look askance at theories that postulate the existence of unknown organs and senses. Matthews (1955*b*) pointed out that well developed homing behaviour is a necessary concomitant of territorial behaviour, sharing the selective advantages of the latter—a point laboured by Ardrey (1966). By a similar token, the sensory mechanism may be expected to be one playing a general part in the life of an animal, particularly as some individuals of a population may be sedentary, others migratory (Lack, 1943/44) and an individual may change its status (Nice, 1933).

A group of theories clearly excluded by the evidence considered in the two previous chapters is that invoking the maintenance of direct sensory contact with home. These would become less effective with increasing distance whereas we have seen that a considerable minimum displacement is needed before navigation comes into play. Such theories may, however, be reviewed briefly, together with additional reasons for their several rejection.

Direct visual perception over immense distances was postulated by Hachet-Souplet (1901) after he had demonstrated the importance of vision in homing Pigeons. Although their eyes are superior in several aspects (p. 142) vision must in any case be limited by the curvature of the earth. Homing birds have frequently been observed to pick up their orientation when flying low (around 200 feet) with the horizon at less than 20 miles. The presence of high mountains, of cloud formations over isolated islands, of sky-glow over pack ice, would increase the effective range of vision, though the latter features would not be characteristic of a *particular* island or coast. Contrariwise, the atmospheric conditions will seldom allow the theoretical range of vision to be achieved. Duchâtel (1901) postulated a sensitivity to infra-red rays which penetrate haze. Wojtusiak (1949) even suggested that by such means birds could navigate visually at night, perceiving the warmer southern regions. Vanderplank (1934) claimed to have demonstrated a visual sensitivity to infra-red rays in owls, but this was denied by Matthews & Matthews (1939), Hecht & Pirenne (1940) and Kopystyńska (1962) for owls, and by Dijkgraaf (1953) for Starlings. From a purely logical viewpoint it is unlikely. Pirenne (1948) draws the analogy of a red hot camera taking a picture of a red hot poker.

A special sensitivity to the physical and chemical constituents of the atmosphere as a means of distant detection of the home area has been frequently suggested since the time of Evans (1795). It is generally looked upon as an extra sense not necessarily allied to that of smell. Various parts of the body have been proposed as its site, such as the air sacs by Fatio (1905). But Hachet-Souplet (1911) and also Oordt & Bols (1929) found that Pigeons homed successfully with these structures collapsed by puncture. Cyon (1900) suggested the nasal cavities, but his own experiments were unconvincing and Watson & Lashley (1915) found the homing of Noddy Terns unimpaired when the cavities were blocked with wax.

Odd theories calling for 'radiations', of an unspecified nature, from the home area frequently crop up in popular literature. Mattingley's (1946) contribution served its purpose by calling forth a devastating counterblast from Thomson (1947). Rhine (1951) and Pratt (1953, 1956) have suggested that some extrasensory means of orientation is the basis of homing. However, even if we could accept their findings as to the reality of telepathy, clairvoyance and psychokinesis in Man, it is difficult to see how such phenomena, if they existed in animals, could be of assistance in homing. Their independence of ordinary considerations of time and space has been strongly emphasized by their advocates, so they could hardly be of use for two-dimensional orientation. Indeed, no suggestions as to the mode of operation have been forthcoming from the parapsychologists, who really became interested in bird navigation only because the known facts had then received no adequate explanation in terms of sensory physiology. This interest has been rejected by Matthews (1956) and there appears to be little activity in this field today. We may also mention here, and dismiss, vague theories of a special 'sense of space' which means nothing and explains less.

To return somewhat closer to sensory reality we may consider theories requiring an animal to remember the outward route through space by registering all the twists and turns to which it was subjected in its box. Spalding (1873) and Darwin (1873) both had this conception in general terms, and Murphy (1873) put forward a plausible mechanical analogy, an early 'biological model'. This was a ball freely suspended from the roof of a railway carriage, reacting to shocks given to it by changes in the latter's direction and velocity. He conceived that '...a machine could be constructed in connexion with a chronometer, for registering the magnitude and direction of all these shocks, with the time at which each occurred; and from these data...the position of the carriage, expressed in terms of distance and direction, might be calculated at any moment... Further it is possible to conceive the apparatus as so integrating its results...that they can be read off, without any calculation being needed.' Without this conception of constant integration the whole hypothesis would quickly be reduced to absurdity since the most exact retracement of the outward path would be required. Detour experiments such as those of Hachet-Souplet (1911), Matthews (1951 b) and Wojtusiak (1960) effectively dispose of any lingering doubt on this point.

Exner (1893) suggested that the semi-circular canals of the inner ear would provide the type of mechanism required to subserve this 'sense'. Meise (1933) added the hypothesis of 'muscle memories', or proprioceptive recordings, in those cases where the outward journey was effected by the animal itself. His proposal is therefore applicable to none of the experimental homing tests. The theory would have to rest, therefore, on labyrinthine recording. On the face of it the whole idea seems quite impossible—that the available apparatus could detect and record every change of direction and acceleration against a background of much greater jolts and jars from the transporting vehicle. However, recent developments in the theory and practice of 'inertial navigation' in aviation and astronautics have shown in man-made vehicles that this is quite a feasible proposition and cannot be dismissed out of hand. We may therefore summarize the main requirements; for more detailed discussion reference may be made to Barlow (1964) as well as the text books on the subject which have blossomed in recent years.

A fundamental requirement is that the inertial system must be capable of continually determining the local vertical (the direction of gravity), despite the movement of the vehicle over the earth. Schuler (1923) pointed out that a plumb line would only give this information if its length were equal to the earth's radius, with the bob at the earth's centre. This is impractical but can be replaced by an oscillatory system having the same period, 84 minutes, as a Schuler pendulum. No evidence is yet forthcoming for such a system in birds. Another basic requirement is that the accelerometers be mounted on a platform stabilized horizontally, with reference to the local vertical. This is because any yaw or pitch will result in spurious accelerations which will be fed into the system. In man-made systems this is usually achieved by means of gyroscopes; other gyroscopes must also keep the accelerometers properly orientated in azimuth relative to the axes of the co-ordinate system employed, e.g. N/S and E/W. Since we can, with a degree of certainty, discount the existence of gyroscopes in a bird, this goes a long way to putting inertial navigation out of count, without the further specifications that a self-contained system must contain an extremely accurate mechanism for measuring time intervals, be continuously precessed to offset the earth's rotation, be corrected for changes in Coriolis Force with latitude (see p. 103) and adjusted to the oblateness of the earth's spheroid. However it now appears that it is possible in principle to build what are known in the jargon as Strapped-Down or Vehicle-Oriented inertial navigation systems. Such a system employs rectilinear sensors, angular sensors and a computor to sort out velocity and heading from changes in orientation of the vehicle itself. The computer and sensing devices must be of much greater complexity than where a stabilized platform is provided by gyroscopes. Barlow (1964) has calculated that the threshold levels for perception of angular and linear acceleration in humans

(0·2/s² and 6 cm/s²) would not permit accurate navigation (5 % errors) for more than a few minutes. Inertial sensing elements with thresholds exceeding the above values by $10^4$ are already used in man-made systems. One must never argue from the limits of human sensitivity to those of other animals, but it appears that inertial navigation is on theoretical grounds a poor starter.

Practical investigation of the possibilities dates from before the realisation of the subtleties of inertial systems, and the glorious theoretical potentialities of adding ever more complex computer and feed-back systems together. However the weight of negative evidence adds further substance to the rejection on theoretical grounds. Thus experiments were made in which birds have been transported under heavy anaesthesia—Pigeons by Exner (1893), Starlings by Kluijver (1935) and Herring Gulls by Griffin (1943). In no case was the homing result poorer than that with untreated birds in comparable circumstances. Another approach is to make the outward journey so complicated that it would be beyond reason for any recording apparatus to cope with it. Rüppell (1936) using Starlings, and Griffin (1940) using Leach's Petrels, rotated their birds on turntables and reported no deleterious effect on homing. However, rotation was limited to part of the journey and/or the releases were not sufficiently far from known territory to make returns by random searching improbable. Matthews (1951 b) therefore carried out a series of tests with Pigeons taken to the release point in a large light-proof drum slowly turning horizontally, *not*, as Ardrey (1966) misinterprets it, like a concrete mixer. The construction was unstable so that changes in the speed and direction of the transporting vehicle produced a momentary slowing of the drum. The outward journey through space was thus remarkably complicated by the irregularly varying rotation, about 1200 revolutions in the longest journey. Nevertheless, in every case the performance of the rotated birds, both in orientation and returns, was just as good as the untreated controls.

It would seem that a final answer would be obtained with the destruction of the labyrinth itself, though such operative treatment is undesirable because of the widespread effects it would have, and only negative results would be acceptable. Normal homing by Pigeons has in fact been reported after blocking of the auditory canals (Casamajor, 1927; Grundlach,

1932); destruction of the tympanic membranes (Hachet-Souplet, 1911); removal of the pars inferior and both rear ampullae (Huizinger, 1935); extirpation or cutting of the horizontal semi-circular canals (Hachet-Souplet, 1911; Sobol, 1930; Huizinger, 1935). The distances involved in tests after recovery from the operation were often not great, though Hachet-Souplet reported good returns from up to 240 miles. Wallraff (1965) after bisecting the horizontal canal in nine experienced Pigeons found that they orientated and homed from three points, 90 to 100 miles SSW, ENE and W just as well as controls. One of the birds subsequently homed from 400 miles SSE. Casamajor (1927) reported impairment of homing after injection of quinine chlorohydrate which, in humans, produces buzzing noises in the ear, and Treat (1947) after decompression to an indicated altitude of 25,000 feet. Both treatments, however, could well have had generally deleterious effects.

Before leaving the ear-apparatus, mention may be made of one more curious hypothesis. In a series of papers Vitali (e.g. 1912) described a small innervated vesicle, the paratympanic organ, in the middle ear. Neither he nor anyone else has been able to assign any sensory function to this organ. A few centuries earlier it would no doubt have been described as the 'seat of the soul'. Vitali believed that it was the site of the 'homing sense'. Benjamins (1926) showed that cauterization of the organ in both ears had no effect on the homing of Pigeons.

Schreiber et al. (1962) rotated Pigeons on a turntable and recorded the electrical discharges from the cerebellum by means of implanted electrodes. They found a marked qualitative difference between the responses obtained from the brains of racing Pigeons and those of unselected town Pigeons; also between the responses from migratory and non-migratory doves. A short time after the rotation stopped, the racing Pigeons and migratory doves showed a series of after-discharges, lacking in many of the other birds. The after-discharges appeared even in deeply curarized birds and certainly expressed cerebellar and not periferal activity. The discharges were constant for individuals and their pattern was inherited, as shown by those of hybrids.

This demonstration of a positive physiological feature associated with homing ability is of considerable interest. The cerebellum is, of course, the centre co-ordinating muscular movements and bodily orientation and is therefore particularly well developed in birds. But while the qualitative difference now shown *might* reflect the existence of a superior vestibular apparatus capable of inertial navigation, it is more likely to indicate a different order of muscular and sensory co-ordination which would provide a more stable 'instrument bed' for observation of external phenomena (chapter 13).

# CHAPTER 8

# Theories of navigation by geophysical 'grids'

The discovery of the distance effect in pigeon orientation (p. 82), good orientation being shown over 50 miles but with a zone of disorientation closer than that, strongly suggests that navigation involves the *comparison* of stimuli. Let us suppose that there is some physical factor, $X$, that varies quantitatively in a regular way across the earth's surface. Let $Y$ be another factor with a gradient at an angle to that of $X$. Then the isolines joining places with equal values of $X$ will cross the isolines of $Y$ to form a 'grid' as in fig. 26. In a perfect 'grid', in which a given isoline of $X$ crosses a given isoline of $Y$ only once, any point will be

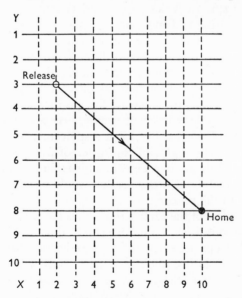

Fig. 26. Diagram of a 'perfect' navigational grid. Two physical factors, $X$ and $Y$, vary in a constant, quantitative fashion across the earth's surface, with their gradients at right angles to one another.

uniquely characterized by two co-ordinates. Thus, in the figure, the home has a value of $X_{10}Y_8$ and the release point of $X_2Y_3$. Clearly the greater the angle at which the isolines intersect the more precisely will a given pair of values characterize a given point—in navigational jargon, the better will be the 'fix'.

A comparison of home and release point values would become easier the farther apart the two points were and, conversely, would require a certain minimum displacement. But another element in bi-coordinate navigation is required if, after such a comparison, the bird is to leave in the direction of home shortly after release. It must 'know' that if factor $X$ is less it has to move in one direction across the grid, if it is more, in the opposite direction; and similarly with factor $Y$. Further, to translate this into practical flying, it must be able to determine the orientation of the *grid* with reference to the surroundings of the release point, much as we 'set' a map by coinciding its northing line with the compass needle.

Ising (1945) made a detailed analysis of the various dynamic consequences of the earth's rotation, and the ways in which they could be used by a bird for navigational purposes. Thorpe & Wilkinson (1946), de Vries (1948) and Wilkinson (1949) made calculations to relate Ising's proposals to the potentialities of the sense organs likely to be concerned. The effects could at most provide only one set of co-ordinates, corresponding to the parallels of latitude, not a complete navigational grid.

First, then, we have centrifugal effects. As a body moves towards the equator, the sideways force is increased, tending to offset the downwards force of gravity. As a result the body apparently loses weight. But for a displacement over 50 miles the change is very slight, of the order of one part in twenty thousand, say 0·02 gm in a Pigeon. This is far less than changes produced by metabolic processes that would be going on during the journey to the release point.

A second effect of the earth's rotation is the production of the so-called Coriolis force. Its nature is best illustrated by Wilkinson's model of a particle situated on the edge of a disk of radius $r$. If the disk is rotating about a vertical axis with an angular velocity $w$, the particle's sideways velocity will be $rw$. If the particle moves towards the axis, say half-way along the radius, its sideways velocity is now $rw/2$. By definition, such a diminution of velocity must have been produced by a sideways force acting in the opposite direction. This is the Coriolis force, of magnitude $2mwv'$, where $m$ is the mass and $v'$ is the velocity with which the particle moves towards the axis. If we imagine the earth's sphere to be made up of a series of concentric disks of diminishing radius, a body moving north over the surface is effectively moving towards the axis as it passes from the edge of one disk to that of another. The rate at which it does so is a function of the bird's velocity $v$, and the latitude $\lambda$, $v' = v \sin \lambda$, giving the Coriolis force as $2mwv \sin \lambda$. Yeagley (1947) suggested the Coriolis force could be detected directly as a sideways force on the flying bird. But this would require the bird to estimate (or hold constant) its ground speed to within 0·2 m.p.h. Similarly its mass must be estimated or held constant to within 2 gm, despite metabolic changes and such incidentals as defaecation. In the unlikely event of these requirements being met the bird would still have to measure a sideways force of less than 1/6000th that of gravity, which would now seem to act at a small angle to its normal direction. This angle would have to be measured to within 0·2 seconds of arc although, since gravity is already supplying the direction of the resultant force, the bird has no other vertical reference from which to measure the angle. A deviation in course of the order of one inch in a mile, in vertical or horizontal planes, would produce spurious Coriolis forces that would mask those due to the earth's rotation. This would seem to be an impossible requirement even though, as we shall see later (p. 153) a bird's *head* is remarkably stable in flight.

The second method of detecting Coriolis force, that originally suggested by Ising, is both more subtle and more feasible since it can be used by a bird at rest and, further, no measure of its total mass is necessary. He considered the effect of the Coriolis force on fluid contained in a ring-shaped tube, and thus of constant mass. If such a tube is held horizontally and then tilted at some constant rate about its east-west axis, the fluid in the northern half of the ring moves towards the earth's axis, that in the southern half away from it. The fluid in each half thus experiences an equal Coriolis force but in a different direction. If the fluid was already flowing round the ring, the result will be a couple tending to turn the ring, analogous to the effect of a magnetic field on a wire loop through which flows an electric current. Ising was able to demonstrate such a rotation in a glass model 20 cm in diameter and with a rate of flow of 30 cm/sec

But the only tubes in an animal with a liquid flowing rapidly through them are the arteries. These are subject to intermittent pressure fluctuations very much greater than the minute lateral pressure the couple would produce.

A second effect of tilting the ring, and the resultant Coriolis forces, would be a streaming of the fluid round the ring. There would then be no requirement for a prior movement of the contained fluid, and the semi-circular canals of the inner ear would be suitable structures for the detection and measurement of the force. On such a scale the amount of energy produced is very small. For a ring 1 cm in diameter and 1 $mm^2$ cross-section, containing fluid of density 1 and viscosity zero, it will, on turning through about 6°, gain a maximum Coriolis energy of $2 \times 10^{-13}$ ergs, only ten times the Brownian agitation energy present in the detecting structure. Moreover, the extra energy is in the liquid and has to be transferred to the detecting sensory element, with a considerable loss of energy. This reduced energy would have to be measured to within $10^{-15}$ ergs to be of use in latitudinal determination, and against a background of the swirling caused by rotation of the ring at right angles to its diameter in its accepted function of analysing postural changes. Thorpe & Wilkinson (1946) showed that there is no tendency for the semi-circular canals of long-distance migrants and proven homers to be larger, relatively or absolutely. Indeed, the diameter is often only half that of Ising's example, with a hundredfold reduction in the energy produced, requiring measurement to within less than 1/1000 of the masking Brownian energy.

We may consider that determination of latitude or direction from the Coriolis force due to the earth's rotation to be most improbable on the biological scale. If birds, kept in constant but irregular motion right up to the moment of release were able to orientate when in flight, the case against the hypothesis would be even more definite. The negative effect of sectioning the semi-circular canal (p. 99) may be called in evidence here too.

Hypotheses that the earth's magnetic field could provide a navigational grid date back to Viguier (1882). He suggested that birds could detect and measure the three components of the field, its intensity, inclination (the angle which a compass needle makes with the horizontal) and declination (the angle between magnetic and geographical north). These vary more or less independently of each other so that their isolines would form a complex grid. This hypothesis has frequently been re-stated, with minor variations, by Thauziés (1910), Stresemann (1935) and Daanje (1936, 1941). Rochon-Duvigneaud & Maurain (1923) objected that measurement of declination re-quires an exact knowledge of geographical north. But we now know that this could be provided by sun- or star-compass. The lack of evidence for any direct reaction to a magnetic field in birds has already been discussed (p. 25). Meyer's (1966 b) negative results were of particular relevance since the changes in magnitude of the experimental field were of the order the bird would have to detect in order to navigate. Further, as described below, powerful magnets have been attached to various parts of the anatomies of birds without affecting their homing ability.

The original concept of a direct sensitivity was replaced by one of indirect sensitivity to the earth's field, and the whole hypothesis resurrected by Yeagley (1947). He suggested that the earth's field could be detected by the flying bird acting as a linear conductor moving through the lines of force of the field. Theoretically this would result in a small potential difference being set up between the two ends of the conductor, though this has not been demonstrated in practice. The induced voltage would require measurement to within one millionth of a volt if it was to be used for navigational purposes. Further, the bird would be required to make an accurate estimate of its ground speed, to within 0·2 m.p.h. An even more cogent objection was raised by Slepian (1948) and others, who indicated that the minute voltages would have to be measured against a back-ground of the far more powerful electrostatic field of the earth, about one volt, and of the fluctuating effects of charged clouds. Stewart's (1957) suggestion, that air friction on a bird's feathers sets up electrostatic forces which react to the earth's field, should not perhaps be taken too seriously. It would certainly be difficult to test experimentally. Talkington's (1967) sugges-tion, that the lymph tubes in the pleats of the *pecten* in the eye could acts as conductors in which an e.m.f. is generated, adds nothing but a theoretical (and far-fetched) alternative. His hypothesis has yet to be published in detail and only very sketchy reports of practical tests are available.

Wilkinson (1949) pointed out that a more satisfactory method of detection would be a conducting loop (such as a semicircular canal) oscillated in the earth's field. On the dynamo principle an alternating current would be induced which is much easier to measure than a potential difference. Also the necessity of knowing the ground speed would be avoided. But the current would have to be measured to within a thousand-millionth part part of an ampere. And once again there would be much more powerful background effects, in this case the physiological currents.

When dealing with biological systems the results of experiments are always more convincing than physical arguments that may be based on false premises. Techniques aimed at disturbing an electro-magnetic apparatus had been reported by Exner (1893) and Griffin (1940). The former passed electric currents through the heads of Pigeons before release, while the latter subjected Leach's Petrels to an intense electro-magnetic field for a few seconds before the beginning of the outward journey from home. In both cases no effects on homing were apparent, but the techniques were less satisfactory than if the bird was subjected to 'interference' during the actual flight. If magnets were fixed rigidly to the head, the additional field would be constant and could be taken into account by the analysing mechanism. It is therefore essential that the magnets should move relative to the bird's body. Yeagley (1947) attached small, powerful magnets to the wings of Pigeons, sewing them on through the metacarpal joints. The fluctuating e.m.f. induced in the bird's body when the wings were beating would swamp any measurement of that induced by the movement of the body through the earth's field. Using ten Pigeons treated in this way Yeagley claimed to have established that the magnets had a strongly deleterious effect on homing. But the difference in performance from that of controls was not statistically significant and several magnets were lost in flight. A similar test carried out by Yeagley in 1945 (Yeagley, 1951), which gave completely negative results, found no mention in his earlier paper. In the meantime Gordon (1948) repeated the test with more adequate numbers and a negative result. It was found later that at least some of his Pigeons had flown over the test course previously, thereby throwing doubt on the results. Matthews (1951 *b*) carried out further tests with wing magnets under really critical conditions and found that they had no effect on initial orientation or on speed of return. Schumacher (1949) suggested that there might be receptors in the wings themselves, cutting the lines of force as they beat. The attachment of magnets to the wings would not then be critical. This implausible point is answered by the negative results of experiments in which the magnets were attached to the heads of Pigeons (Casamajor, 1927), Storks (Wodzicki *et al.* 1939) and Swallows (Bochenski *et al.* 1960), and of other negative tests in which larger and more powerful magnets were suspended from the necks of Pigeons (Matthews, 1951b) or the legs of gulls (Matthews, 1952a), so that they oscillated freely in flight.

The modified hypothesis of magnetic navigation would provide only one set of isolines, one co-ordinate of a grid. To provide the other set, Yeagley proposed that birds were also able to detect and measure the Coriolis force due to the earth's rotation, which we have already seen is unlikely. Since the two systems were based respectively on the magnetic and geographical poles, a given magnetic isoline would cross the same Coriolis isoline at least twice, producing two 'conjugate' points indistinguishable from each other by a bird using the proposed type of navigation. If Pigeons trained to home to one conjugate point were released near the other, they should home to the latter. Yeagley tested this conception by training Pigeons to a conspicuous mobile loft in Pennsylvania. To ensure that the birds would home to the loft in different surrounding, the structure was moved bodily every day, over several miles. When training was complete the entire set-up was transported 1400 miles to the conjugate point in Nebraska. In a number of cases at least the birds were allowed to fly freely around the new loft site. They were then released, usually at distances of 50 to 80 miles, sometimes as little as 25 miles, in small groups. From 459 Pigeons so released when one or two lofts were in position near the conjugate point, only eight regained a loft. Five of these were from a single release and could well have remained together. This was less than one would expect on a random radial scatter, and much less than would be achieved by random exploration (p. 67). No ground observations seem to have been made of the initial direction taken by the birds. Ten groups of Pigeons were followed by a light aircraft for more than 25 miles from the release point east of the conjugate point, and these did show a westerly trend. But emphasis cannot be laid on results obtained from one direction.

The bulk of Yeagley's analysis depends on reports of the positions at which Pigeons were subsequently recovered, both in experiments with the loft(s) present and in others where the birds were simply launched into the void of Nebraska. He split the 'flight lines' into two components, and then summed these components algebraically to obtain an 'average flight vector'.

A more reasonable approach would be to determine whether the 'flight lines' showed any significant tendency to concentrate about the bearing of the conjugate point, to determine their average deviation. Griffin (1952c) calculates that the data in Yeagley's 1947 paper give an average deviation of 64°, which is hardly impressive (illustrations of these scatters are given in Kramer (1948) and Matthews (1951a)). For his 1951 paper the value is 80°, and fig. 27 shows clearly the essentially random scatter of recovery points. The data in Yeagley's papers are insufficient for any assessment of possible topographical biases to be made, and there is no precise information on winds. Odum (1948) gives instances, from a general study of weather maps, in which the results could be interpreted in terms of wind drift.

There should at least be a demonstrable tendency for the recoveries to be closer to the conjugate point than when they were released. Considering all the 175 recoveries of trained birds released in Nebraska which did not remain at the release point, we find that on the average they were released 70 miles from the conjugate point but were recovered 105 miles from it. Yeagley's hypothesis is therefore unacceptable not only because of its theoretical 'impossibility', but also because the massive field experiments have produced negative results.

Attempts such as those of Talkington (1967) and Graue (1965b) to relate the homing tracks of Pigeons to quirks in the magnetic field are not yet convincing. Reports that birds are affected by electrical disturbances, natural and artificial, are generally considered to have a bearing on the theory of magnetic navigation. There is a considerable conflict of evidence. Thauziés (1910) purported to show that thunderstorms produced poor homing results, but Gibault (1928) gave another analysis of the data showing no such correlation. Yeagley (1951) claimed a general decline in the speed of pigeon-race winners with increasing sunspot activity on the day *before* the race, but made no

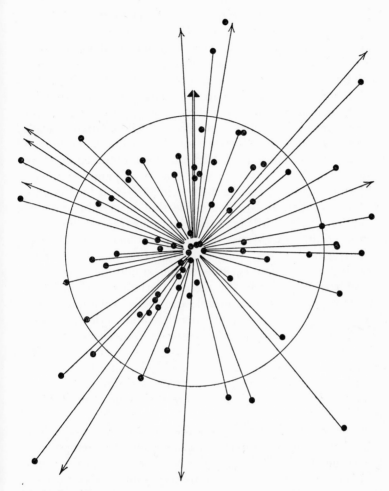

Fig. 27. Recoveries of Pigeons released near the Magnetic/Coriolis 'conjugate point' in Nebraska. Radiating arrows refer to single recoveries outside the limits of the diagram. The large vertical arrow indicates the mean bearing and maximum distance of the conjugate point in 8 experiments. The circle has a radius of 100 miles. (After Yeagley, 1951.)

check on other possible factors operating during the races, such as wind direction and force.

Casamajor (1927), Brown (1939), Yeagley (1947) and Wojtusiak (1960) report interference with orientation by non-pulsed radio transmission of various frequencies. But Gibault (1938), Casamajor (1930) and Meyer (1938) denied any such effect, while Kramer (1951b) in a series of careful training experiments was unable to obtain a reaction to short-wave emissions. Wojtusiak (1960) reported that Swallows 'equipped with aluminium helmets reflecting short radio waves' came back faster and in a higher percentage than control birds. However laboratory tests failed to show any change in behaviour under the influence of such waves. Reactions to *pulsed* radio, or radar, transmissions have been reported by Poor (1946), Yeagley (1947), Drost (1949), Knoor (1954), Hochbaum (1955) and Tanner (1966). Negative results have been obtained by Hardy (1951), Matthews (1951 b), Busnel et al. (1956) and Eastwood & Rider (1964)—who reviewed the evidence—and by Houghton & Laird (1967). It is still possible that in certain undefined conditions, flying birds may react to radar transmissions vastly exceeding in power any natural phenomena of this type. Schwartzkopff (1950) suggested that this might be due to electrical stimulation because of the amplitude modulation, through rectification in the tissues. Certain radar equipment can be made to produce audible sensations in the human subject at suitable pulse-rate frequencies, and Barlow, Kohn & Walsh (1947) report visual sensations from electro-magnetic stimulation. The gross effects of powerful radar transmissions have been used to cook chickens, and to kill insects. Therefore, even if definite evidence that radar transmission affect birds is obtained, it will be no indication of their type of navigational equipment.

Kramer (1959), in restricting the use of the sun to the determination of compass direction, was driven to contemplate unknown geophysical factors as the basis for the 'map' or 'grid' which gave the position of the release point relative to that of home. He was unable to make any positive suggestions on this score but drew attention to a number of anomalies which had been revealed in the orientation and homing of Pigeons, and which might give an indication of the nature of the forces involved. Directional tendencies we have dealt with earlier. There were also found deviations from the home direction characteristic of certain release points, although the deviation did not necessarily remain constant through the season, or even in the course of a single day (Wallraff, 1959b). The effect of landscape features, attractive perhaps because of similarities with those at home, and perceived or not according to visibility conditions, cannot be ruled out here. Another rather mysterious finding by Kramer (1954) and Kramer & St Paul (1956a) was that homing in Germany was poor in the winter months, even in fair weather, though not directly correlated with temperature. This was not so in North Carolina, nor in England. The demonstration by Wallraff (1960a) of a statistical correlation between homing performance and barometric changes at 20,000 to 40,000 feet, but not at the bird's own level, comes into the same general category. General depressive effects of weather syndromes are known in ourselves and may well be concerned in Pigeon psychophysiology as well. Indeed as the bulk of the evidence for these odd *Effekts* comes from releases at short distances, within the zone of disorientation, their relevance to the problem of navigational ability is, at best, marginal.

CHAPTER 9

# Theories of navigation by astronomical 'grids'

Although human navigators have for centuries been obtaining their positions in unknown areas from the co-ordinates of heavenly bodies, it is only in the last few decades that the possibilities of birds doing likewise have been seriously considered. Ising (1945), seeking a second co-ordinate to form a 'grid' with his proposed Coriolis force isolines, postulated a determination of longitude displacement by measurement of time differences in sunrise or sunset. He recognized the effect of latitude changes on sunrise/sunset times, and suggested that the noon bearing of the sun was the only suitable one in such circumstances. But having, as he thought, provided means of determining latitude, he did not point out that the altitude of the sun at its noon position also gives an indication of latitude. This was done by Wilkinson (1949) who made the further, essential suggestion that 'actual observation of the sun at noon is not necessary, occasional glimpses would combine with a time sense to enable its course to be constructed'. Matthews (1951 *a, b*) put forward a synthesis of these ideas as a working hypothesis of complete sun navigation. Pennycuick (1960*a*) offered a variant on this theme, and meanwhile Sauer (1957) proposed complete navigation from the star co-ordinates.

Before considering these hypotheses and detailed evidence on their plausibility, we may first review the general evidence that astronomical navigation may afford the bases of the homing ability examined in chapters 5 and 6.

Matthews (1952*a*) found that the homeward orientation of Lesser Black-backed Gulls in sunny weather deteriorated markedly in conditions of heavy cloud (fig. 28). Similarly Matthews (1953*c*) showed that Manx Shearwaters released in such conditions showed no signs of homeward orientation (fig. 29). Instead the birds scattered in all directions or drifted down-

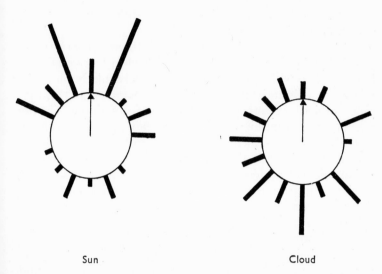

Sun                                                    Cloud

Fig. 28. Lesser Black-backed Gulls showing a homeward orientation of vanishing points when released in sunny conditions, but a disoriented scatter under heavy cloud. (After Matthews, 1952 a.)

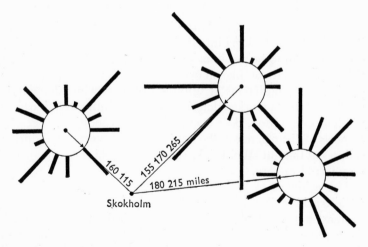

Skokholm

Fig. 29. Manx Shearwaters showing a disoriented scatter of vanishing points when released under heavy cloud. Compare with fig. 25. (After Matthews, 1953 c, 1955 a.)

wind. Matthews (1964) showed that the same *individual* Shear-waters that orientated well in sunny conditions did badly with overcast and vice versa.

In the case of homing Pigeons, Matthews (1951*b*, 1953*a*) found that birds which had shown homeward orientation no longer did so with overcast conditions, particularly on critical releases at a distance off the training line. Kramer (1953*a*, 1957) reported similar disorientations in overcast conditions. The time taken to reach vanishing point was also fairly consistently longer (see also Schmidt-Koenig, 1958; Wallraff, 1960*b*). Indeed it became so axiomatic that good orientation and overcast conditions did not go together that most subsequent observations on cloud effects were accidental and consequential to a deterioration of forecast weather conditions. One exception to the general rule, that reported by Hitchcock (1955), should not receive overmuch stress since only two groups of Pigeons were observed to fly precisely in the home direction under thick overcast and they were lost behind trees at a quarter mile distant. Schmidt-Koenig (1965*a*) has objected to the general inference, that observation of the sun is a necessary part of bird navigation, on the grounds that cases of ill directed and random orientations have been recorded in sunny conditions. His logic is at fault in arguing from the exception to the general rather than vice versa. Moreover a great part of the poor orientations in sunny conditions took place at short distances, where, owing to the 'distance effect' he has himself fully confirmed, true navigation does not seem to be possible. Wallraff (1966*c*) has re-examined the German data on pigeon orientation under overcast and has carried out experiments at suitable distances which have fully confirmed that a marked deterioration, not distinguishable from random orientation, does occur in cloudy conditions (fig. 30). Reports of disturbance in the homing of Pigeons released during eclipses of the sun (Wojtusiak, 1960; Bochenski *et al.* 1960) may be referred to here.

The evidence of Walcott & Michener (1967), obtained by radio-tracking, is quite straightforward. Pigeons released under overcast soon landed and did not fly until the sun reappeared. Even more striking, Pigeons starting in sunny conditions and then meeting sun-obscuring cloud, would land and wait until the cloud had passed. These observations are strong evidence against the contention by Kramer (1953*a*) that the breakdown of homing orientation under cloud was simply due to the sun-compass being unavailable to 'set' his 'map' of co-ordinates based on unknown factors (p. 102). There is plenty of evidence (chapter 3) that a simple compass direction can be derived from the sun through cloud of moderate thickness, and that such courses once initiated can be maintained for a time under thick overcast, presumably by transference to visual reference points in the landscape below. Matthews (1952a, 1953a, c) had earlier found that true homeward orientation (as opposed to learned

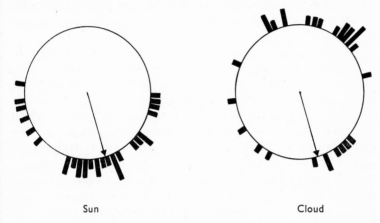

Sun                                    Cloud

Fig. 30. Pigeons, carefully matched for experience, and released on six occasions at a point 90 miles from home, show homeward orientation of vanishing points under sunny conditions, disoriented scatter under heavy cloud. (After Wallraff, 1966c.)

compass orientation) in Gulls, Shearwaters and Pigeons was disrupted by cloud conditions (ca. 7/8ths) which permitted short glimpses of the sun sufficient to bring the simple sun-compass into action. We saw (p. 92) that there is some doubt whether the Golden Plovers of Sauer (1963) in their treadle-cage were showing true homeward orientation. But in this connection it is interesting to note two other of his findings. Under heavy overcast the birds showed no orientation. When they could see the sun's position, but not (because of clouds or the walls of the apparatus) the sun itself, a north-east orientation resulted. This could be interpreted as a northward compass orientation, shifted because the birds were being kept on the Hawaiian time schedule, more than 2 hours behind San Francisco time.

Schmidt-Koenig (1961*b*) used his demonstration of a simple time-compensated sun-compass in Pigeons to support the contention that this was the sun's only function in homing. However we have seen (chapter 6) that the bulk of the releases were at short distances where true homeward orientation is not found, and also that the influence of 'nonsense' tendencies was not then appreciated. The fact that Schmidt-Koenig (1958) had a few, accidental, overcast releases which still showed a difference between time-shifted experimentals and the controls strengthens the interpretation that he was *only* dealing with initial compass orientation possible under any but the thickest overcast.

Whereas swift homing flights occur after sunny releases in Shearwaters and Pigeons, they are much less frequent after cloudy releases. Thus of 128 Manx Shearwaters returning after being released mid-May to mid-June in cloudy conditions, only 6 % came back on the first night, as against 15 % of 131 which were released in sunny conditions; the proportions returning late (after the 10th night) were 18 and 7 % respectively; losses amounted to 25 % against 14 % of the totals released (Matthews, 1953 *c*). A similar general relation in the homing times of Pigeons was also found (Matthews, 1953 *a*). There were exceptions, however, and of course it is unlikely that conditions at release would be maintained over the whole track; the birds might well have an opportunity to establish their orientation not long after they are out of sight of the liberator. Kramer, St Paul & Wallraff (1958) likewise reported some overcast releases which gave poor homing performances, others which did not. Matthews (1964) found with his Shearwaters that the effect of an initial cloudy start on the homing performance became less marked as the birds gained experience of transportation and release. Presumably they were then less likely to be unduly disturbed if unable to determine the home bearing at release. Matthews (1955*b*) reported an instance in which twenty-five highly experienced Pigeons were released at the beginning of a 10-day period in which the whole of Britain was continuously wrapped in gloom and overcast. Only two birds returned on the day of release, two on the second, six on the third, two on the fourth, one on the sixth, one on the eighth and six on the tenth, when the weather improved. Such a result does suggest the absence of navigation, with homing merely by random wandering. It certainly does not suggest the existence of an unknown 'grid' over which the bird could 'feel' its way, by a type of kinesis, along the resultant of physical gradients. This should be possible even if the compass of Kramer's map-and-compass hypothesis were eliminated by cloud conditions at release.

While it thus appears likely that bi-coordinate navigation by birds may have something to do with the interpretation of the sun's position, there is much less evidence to suggest that the interpretation of the stars' positions at night has a similar role. Indeed we have little evidence that birds can orientate towards home from an unknown release point at night. We have seen (p. 48) that the planetarium evidence, provided by Sauer, that warblers detect and react to changes in the star patterns that could be due to longitudinal shift, is at best equivocal. Sauer (1957) further claimed that when he tilted the axis of the projector, so throwing the star patterns appropriate to lower latitudes on the dome, the mean direction of a Lesser Whitethroat changed progressively from SSE to S, much as one would expect if the bird (reared in captivity) was in fact migrating from Germany through the Levant and then down to Africa. Wallraff (1960 b) showed that the change was not nearly so marked as had been claimed and that it was just as acceptable to state that one general direction was maintained. Wallraff also disputed, with good reason, Sauer's claim that migration activity declined when a low latitude sky (appropriate to the end of the journey) was shown. Meanwhile Sauer & Sauer (1960) in further planetarium experiments with a Garden Warbler and a Blackcap had found no effect of changing latitude. They had also (1959) made the ambitious experiment of shipping warblers in the autumn to South West Africa and observing their behaviour there, under natural night skies. Only one bird, a Whitethroat, was fully active in the round cage. Despite the fact that it was exposed at the extreme southern limit of this species, it headed strongly southwards. Such activity as was shown by three Garden Warblers and one Lesser Whitethroat (this 2000 miles south of its normal winter quarters) was also directed southwards. The orientation was certainly no longer appropriate, but the Sauers still claimed that the southern skies were interpreted in terms of a change in latitude. This was because six Blackcaps and two Lesser Whitethroats showed undirected migration activity when caged indoors but became inactive when exposed to the night sky. Wallraff again queried this interpretation, pointing out that indoors the birds were together in the same room (which is known to stimulate activity) but outside were isolated each in its own round cage. Another question which can be raised is why, if southern skies in autumn inhibit migration, do northern skies at the breeding latitude in spring stimulate strong northerly orientation (p. 48)—or, for that matter, produce 'conflict' in the planetarium.

Other evidence (see also p. 93) from caged birds suggesting an appreciation of position by reference to the star pattern is provided by Hamilton (1962 c). A single Bobolink, shipped from North Dakota, on its first night of exposure to the natural San Francisco skies, showed a north-easterly tendency (i.e. in the direction of home); on the third night the tendency was SSE (the normal migration direction from North Dakota). The scatter of vectors was wide and the case is of interest only because the bird then escaped and was retrapped the following summer in the place from which it was originally taken.

Thauziés (1913), Lincoln (1927), Clarke (1933), Nicol (1945) and St Paul (1962) have all reported night homing by Pigeons but only over distances up to 25 miles, after intensive training and with additional guides such as lamps over the loft. An intimate knowledge of local topography plus directional training would seem sufficient explanation, even for the night-races in Hawaii mentioned in St Paul's paper. Pigeons are essentially diurnal animals and thus basically unlikely to practice navigation by the stars. Too much stress should not be laid on the case of a single Purple Martin which homed 234 miles through the night (Southern, 1959). Wojtusiak *et al.* (1937) released Swallows at night but had no returns during the hours of darkness. Matthews (1963 c) made an experiment with Manx Shearwaters which habitually fly at night, at least in the neighbourhood of the nesting colony. Of six birds released just off the home island, five were back in their burrows the same night. Of twenty birds released after dark well inland, in an area unknown to them, 60 miles from home, not one got back that night. Eleven returned the following night, showing that the urge to return had not been lost. Now that methods of attaching lights to birds have been developed it would be interesting to find whether any initial orientation develops at night. St Paul (*loc. cit.*), who first used this method, in 1956, found none in her Pigeons. Nor did Heyland (1965) with Mallard and Pintail ducks taken from their nests. But the distances involved were small, 25 miles under clear skies.

Despite this lack of practical evidence for navigation by the stars, it is convenient to consider the theoretical bases for astronomical navigation by first considering the situation at night.

The Pole Star, Polaris, is very close to the axis of 'rotation' of the celestial sphere, so it stands in a virtually constant relation to an observer on the earth's surface. It is due north and its height above the horizon (altitude) is proportional to the observer's latitude. All the other stars describe westward arcs across the sky and only the highest point (culmination) of such an arc has an altitude which is a constant function of latitude. Culmination takes place across the north/south meridian for the observer.

Thus observation of the Pole Star, or of another star at culmination, could give a bird both its compass directions and an indication of whether it was north or south of home. Indeed since a unique band of stars culminate successively *overhead* at home, a bird which sought to get back under that band would arrive at its home latitude and then only have the choice between west and east to make its final search. This is believed to have been the method used by the ancient Polynesians on their vast Pacific voyages (if these were not accidental as Sharpe (1963) would have us believe).

To fix position at release relative to home the second co-ordinate of a true navigational grid must be introduced, time. Longitude and time are, in effect, one and the same thing, based on the regular rotation of the earth. Longitude cannot be measured independently of time. Human navigators proceed by measuring the altitude of *two* stars in quick succession, time being given by a preset chronometer or, more recently, checked by radio signal. At any one time a star will be overhead at one point on the earth and have a decreasing altitude on a series of circles with increasing radii centred on this point. Thus a given altitude means that the bird must be somewhere on the appropriate 'circle of position' or 'Sumner circle'. Measuring the altitude of a second star would place the bird somewhere on a second Sumner circle and if these circles overlap the observation point must be in one of only two places (fig. 31). Even this amount of ambiguity would be avoided if a timed altitude were measured for yet a third star. The three Sumner circles cross each other at a unique point or form a small 'triangle of position' within which the observer must be.

Of course it is not suggested that a bird would be plotting its position as a human navigator would do, by mathematical calculation or on charts, using an almanac giving precalculated

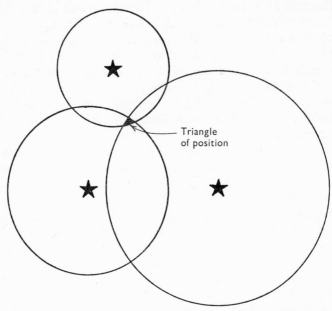

Triangle of position

Fig. 31. Sumner circles. All points on one circle are the same distance from the geographical position of its central star. There it is vertically overhead, so points on the circle all have one value for its altitude. Altitudes for two stars have identical values, at the same time, at only two points, where the circles cross. The altitude of a third star at the same time gives a unique set of values which could only be obtained in the (black) 'triangle of position'. For simplicity the earth's surface is shown flat.

data on star altitudes. More plausibly they would be *comparing* the tilt and degree of rotation of parts of the star sphere with those they would expect to see at home at the same time. They would then move so as to counteract the change, e.g. northwards if the pole of the sphere had been raised, westwards if the stars had been advanced further across the sky. It is interesting to note that Randic (1956), using the original idea of Marcuse, describes a device for the rapid determination of position without calculation. This in effect detects the apparent shift of the star pattern in two co-ordinates, using the zenith (the point vertically overhead) as reference point.

Until definite evidence of night navigation is forthcoming we should rather consider astronavigation as the possible basis of the undoubted faculty for position finding by day. By this is not meant literally navigation by a plurality of stars, though the possibility of birds seeing stars during the day should not be discounted out of hand. We ourselves can, expecially if aided by devices such as pin-hole cameras or their equivalents, pick out bright stars if we know exactly where to look. Gebel *et al.* (1960) have described scanning devices for daylight star observation. However, birds' eyes are not especially adapted in these ways and it is unlikely that they see more than two heavenly objects by day, the moon and the sun. And since homeward orientation and swift return flights are not confined to times when the moon is above the horizon by day, we are left with the sun to consider as the provider of astronavigational fixes.

# CHAPTER 10

# Theories of navigation by a 'grid' derived from the sun's co-ordinates

The hypothesis put forward by Matthews (1951 *a*, *b*, 1953 *a*) in effect considered the sun as the star it is, describing a westward arc across the sky and culminating on the north/south meridian. The time of culmination is local noon and the same for all points on one parallel of longitude. The angle at which the sun arc is inclined to the horizontal is constant for all points on one parallel of latitude. But as the earth is tilted relative to, and in orbit round the sun, the sun arc appears to rise and fall with the seasons so that the altitude at culmination is not constant with latitude. The change is slow (except at the equinoxes, see p. 132) and as the culmination point also gives a measure of local time, Matthews proposed that this point was the one with which comparisons with the home value were made. Further north than home the culmination point is lower, further south higher than at home. Further east the culmination point will be reached before, further west after it would have been at home (fig. 32).

So far the bird is being considered to do what the human navigator does when he determines the maximum (noon) altitude of the sun and checks home time against his chronometer. The need for an accurate chronometer is paramount in any scheme of astronomical navigation—time is longitude, longitude is time—and is no longer such an astonishing concept in a biological system as it was. The evidence for time keeping mechanisms of the necessary accuracy is discussed in the next chapters. For the moment we may consider the difficulty that arises from the fact that birds show evidence of navigation when released other than around noon. This problem also arises in human navigation, when the sun may be obscured at noon. Here it is tackled by the Double Altitude solution, by taking altitudes of the sun when it is visible, preferably on either side of expected noon and, in effect, calculating the noon altitude (May, 1950). Matthews postulated that the bird did something rather similar, observed the movement of the sun along a small portion of its arc and then, by extrapolation, guestimated the culmination point. He stressed that there was no need to consider the bird to do other than react to changes in the sun arc by seeking to restore the norm. Thus, if the culmination point were too low, the bird would fly towards it; if too high, away

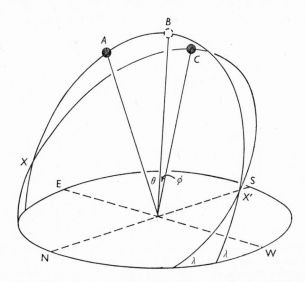

Fig. 32. Perspective diagram of the changes in sun arc consequent on a move to the south and west, at noon (home time). The altitude of the highest point of the arc (B) is greater (by $\phi$) than at home (C). The inclination of the arc ($\lambda$) is also greater. The observed sun (A) has not moved so far round its arc (by $\theta$) as it would have done at home. Note the crossing over of the arcs at two points, X, X'. The earth's surface is shown flat with the horizon ESWN.

from it. If the sun had not moved far enough round its arc the bird would seek to correct this by flying in the direction which adds its movement to that of the sun; if the sun had moved too far, it would fly in the direction which subtracts its movement from that of the sun.

Kramer (1955, 1957) attacked the hypothesis, visualizing the task of the bird as being similar to that of a human navigator drifting in a boat in the open ocean and forbidden the use of a sextant. The bird, as a highly-evolved flying machine, is in no wise so handicapped, as we shall see later. Kramer further considered the sun path to be represented by the line obtained by plotting sun altitude against time (fig. 33*a*). This no more represents the sun path than does the graph of azimuth against time discussed earlier (chapter 3, fig. 10). Later he more properly considered altitude plotted against azimuth, but as this was drawn on a flat surface (fig. 33*b*) instead of a globe, distortions were inevitable. Therefore his argument that such a curve was not open to construction by the extrapolation of a small part of its length, while true, was irrelevant.

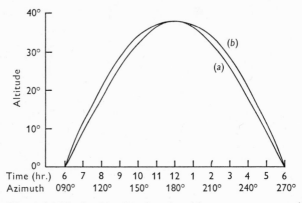

Fig. 33. The sun's altitude plotted against time (*a*) no more represents its path through the sky than do the curves of azimuth against time in fig. 10. Plotting altitude against azimuth (*b*) still does not represent the true path owing to the distortion inherent in displaying spherical geometry on a flat surface.

More importantly Kramer (1957) completely rejected the possibility of the bird being able to detect and measure changes of sun azimuth, on the grounds that parallax would swamp the results obtained by using horizon features as reference points. Pennycuick (1960*a*) took up this point and, while not rejecting out of hand measurement of azimuth changes when over land, roundly declared it to be not physically possible over the sea. However Pumphrey (1960) stated 'Assertions about the physical impossibility of animal activities have so often proved false that they should be closely scrutinized, and Pennycuick's argument appears to be fallacious. What the bird needs is not a fixed object (in parenthesis it may be pointed out that a compass needle is not a fixed object) but an indication of a direction—any direction—in the horizontal plane lasting long enough (not more than a few minutes on Matthews' hypothesis) to allow the change of azimuth of the celestial object to be measured'. He went on to point out that wave patterns on the sea would provide the necessary stable indication of direction, a point which Pennycuick (1960 b), with some reservations, conceded.

In the course of his criticism of the extrapolation hypothesis Kramer (1957) put forward an alternative hypothesis, not dissimilar to one outlined by Griffin (1955), crediting the bird with a time scale sufficiently accurate to differentiate between very small fragments of sun arcs as observed from various points of the globe. This he immediately dismissed. However Pennycuick (1960a) developed the hypothesis in full theoretical detail. It differed in two main respects from that put forward by Matthews. First, measurements and comparisons with remembered home values were to be confined to the time of the observations, there was no extrapolation. Secondly, measurements were only to be made of the vertical components of the sun's movement, its altitude at the time of observation and the rate at which that altitude was changing. In other words, as in Matthews' hypothesis, the bird was required to determine where the sun was on its arc and what the slope of that arc was at that time, functions of the inclination of the arc to the horizontal (which we have seen is constant for any given latitude).

When the lines of equal altitude and of equal rate-of-change are constructed they form a symmetrical grid centred about the sun position and beneath which, as it were, the earth rotates anti-clockwise (fig. 34). The sun stands to this grid in the same relation as does the north pole to the latitude/longitude grid. But, again, there is no need for the bird to translate its information gained from the sun position to fixed co-ordinates on the earth. If the sun altitude is too high it flies away from the sun, if too low, towards it; if the rate of change of altitude is too great it flies left of the up-sun position, if too small, to the right. The angles laid off will be proportional to the differences and will vectorize to give the homeward direction. In checking this formula against fig. 34 it is necessary to remember that compass directions on the earth are in relation to the latitude/longitude grid, *not* to the sun grid under which it rotates.

From the spacing of the isolines, particularly those of rate-of-change, it will be apparent that a given error of measurement (or comparison) will be more serious, the lower the latitude and the earlier in the day that it is made. Pennycuick gives detailed curves for different latitudes, times and seasons. It so happens that Matthews (1955 *b*) found a small, but significant, increase

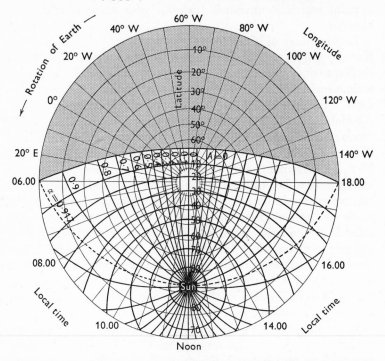

Fig. 34. The grid for sun altitude/rate of change of altitude. The projection looks down on the northern hemisphere the north pole being in the centre. The Earth rotates anti-clockwise under the grid of isolines of sun altitude (*A*) and rate of change (α) centred on the Sun. The line *A* = 0 represents sunrise; this is occurring farther west the higher the latitude (e.g. at 20° E, 20° N and also at 20° W, 60° N). The grid is shown in its midsummer position relative to the Earth, so the sun does not set above $67\frac{1}{2}$° N. (After Pennycuick, 1960 *a*.)

in gross errors in the orientation of Pigeons released up to 0900 hours (19 % of 232) as compared with those released within 3 hours of noon (11 % of 320). This could, however, equally well be adduced (as it then was) to the greater amount of extrapolation needed earlier in the day. Wallraff (1959 *b*) has also provided some limited data on variations in orientation accuracy with time of day.

The isolines of instantaneous altitude are, of course, Sumner circles (distorted by the projection on a flat surface) about the sun position. It will be seen that they generally cross the lines of longitude at an angle. If therefore there were some method of determining local time (the same thing as longitude) a fix would be possible using the instantaneous altitude. However, as we have seen, the only way of finding local time is by determining the time of culmination on the sun arc, and if we know that it would be better to use the culmination altitude to fix position in any case. Inspecting fig. 34 it will be seen that in the morning and evening in the medium latitudes (around 50°) there are areas of the grid where the instantaneous altitude isolines parallel the longitude ( = time) meridians. These times correspond to the cross-over points of the two sun arcs shown in three dimensional representation in fig. 32. Places far apart in latitude would then have the same instantaneous sun altitude at the same time. Kramer (1953 a) released groups of Pigeons before, at and after the morning cross-over point 200 miles south of the loft. There was no apparent difference in orientation with the time of release, all groups departing northwards. This was thought to be conclusive evidence against the use of instantaneous sun altitude to gain information about latitude displacement. The later revelation of a northward 'nonsense' tendency in Wilhelmshaven Pigeons (p. 78) raises doubt as to whether these birds were in fact initially involved in determining their position relative to home. The experiment would certainly be worth repeating. Another glance at fig. 32 shows not only that there is an absence of latitudinal information at the cross-over point, but prior to that the sun is actually *lower* at the more southerly latitude than at home. This reversal of the usual situation is easily sorted out if the arcs are extrapolated either to culmination or back to the intersection with the horizon. It is also taken care of if the rate of change of altitude is measured and the rule of thumb reaction applied with the sun successively NE, E and SE i.e.:

before cross-over:
    sun too low, fly towards it          (NE) ⎫ N
    rate of change too great, fly left   (NW)⎭

at cross-over:
    sun same altitude, no information   (—) ⎫ N
    rate of change too great, fly left   (N) ⎭

after cross-over:

| | |
| --- | --- |
| sun too high, fly away from it | (NW) |
| rate of change too great, fly left | (NE) |

N

Tunmore (1960) proposed that the bird concerned itself only with the instantaneous sun altitude and compared it with that remembered as obtaining at home at the same (home) time, using the first only of Pennycuick's rules of thumb, i.e. sun too high, fly away from it; sun too low, fly towards it. This would not head the bird directly towards home but send it off on a

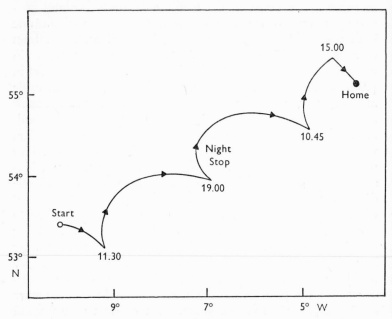

Fig. 35. The curved path a bird would follow if it homed by attempting only to restore the sun altitude to the appropriate value for the time according to its internal 'chronometer'. In this example a Pigeon released 280 miles from home and flying at 19 m.p.h. would take two days to home. The final sharp turn is where it enters the area of known landmarks. (After Tunmore, 1960.)

complicated series of spiral curves varying in shape according to time, season, latitude, speed, wind, etc. One of his examples is shown in fig. 35. Despite the varying directionality of the flight, the bird would eventually be brought close enough to home for it to complete the journey by landmarks. The sharp inflections in the bird's track occur when the Sumner circle of altitude, on which home is at that instant, sweeps past with the earth's rotation. The sun which was before too high is now too low and the direction of flight with regard to the sun position is switched accordingly. While this form of limited navigation would not give the observed homeward orientation and swift homing flights that have been recorded, it could account for some of the slower returns. It might be that some species, or perhaps some individuals of some species, are equipped in this minimal way, the more advanced homers being capable of bi-coordinate sun-navigation of one or other of the kinds under discussion.

The grid of isolines in fig. 34 is symmetrical about the sun position. So for every point north of the (dotted) circle of greatest rate of change of altitude there will be one point to the south with identical co-ordinates. The possibility of am-biguity north and south of the sun path is, of course, inherent in all types of sun navigation as well as in sun-compass orienta-tion (p. 36). When the sun declines from the summer solstice position shown in fig. 34 to the equinox position, the circle of greatest rate of change coincides with the earth's equator (and has the maximum possible rate of 1° altitude change per 1° hour angle = 15 minutes). The cross-over points are then on the horizon and the sun rises at 06.00 and sets at 18.00 (local time) in all latitudes. The further decline of the sun to the winter solstice takes the circle of greatest rate of change below the earth's equator. Field tests to exploit these north/south ambiguities will afford some of the best opportunities to test the reality and nature of sun navigation.

The grid of isolines being symmetrical on either side of the noon position, there are points with identical co-ordinates in the morning and the afternoon. As Schmidt-Koenig (1961 a) pointed out, attention was not drawn to this ambiguity in the original formulation of Pennycuick's hypothesis. However, Pennycuick (1961) saw no reason why the bird should not dis-tinguish rate of change upwards from rate of change downwards, which would remove the ambiguity. Yet this does disturb the essential simplicity of the original statement, for a third para-meter, that of *direction*, has been introduced.

While it is always risky to argue from what the human visualises to what the bird visualises, the procedure can be in-structive. The first thing we notice about a moving object is the direction of movement, and this we do long before its velocity can be measured and vectorized. If we accept that the bird appreciates and reacts to movement up and down, it is difficult to see why it should ignore movement across, particularly when the vertical component is small, in the hours near noon. By allowing the distinction between clockwise and anti-clockwise movement, this opens up the possibility of resolving some of the ambiguities north and south of the sun path. Much of the work on the sun-compass orientation suggested that the birds reacted to the sun position in azimuth only, ignoring or rejecting the information available on sun altitude. However it now (p. 38) appears that factors correlated with altitude are taken into account. By the same token one must be wary of asserting that azimuth information is ignored or rejected in the homing process, particularly as detection of sun movement in azimuth by the flying bird is not, in principle, impossible.

Both versions of the hypothesis require very fine analysis of the sun's movement and the feasibility of such measurements is discussed later. Both require the comparison of observed values with remembered values of the home arc. Matthews, by introducing the added complication of extrapolation, required the bird to remember only one point on the home arc, the culmination point. This is also the point when the difference between arcs is greatest. Pennycuick, by dispensing with extrapolation, required the bird to remember the whole of the home arc so that comparison can be made at any instant, corresponding with that of observation. In the morning and afternoon this will be where the separation of the arcs is minimal (fig. 32). Advantages and disadvantages appear to be pretty evenly balanced. Pennycuick is sanguine about the memory requirement, pointing out that three numbers are sufficient to specify the sun's motion completely at any given point. But one can also suggest that the bird is more likely to remember the shape of the arc from the envelope of tangents to points along it, not the co-ordinates of those points. Indeed there are only three points on the sun arc which could be expected to have special significance in the home situation, sunrise, noon and sunset. Of these noon is singled out both by reasons of symmetry and because the azimuth, due south, is constant throughout the year and forms the basis of the home compass-rose and of local time. In the latter connection the report of Vaugien & Vaugien (1963) is of interest. House Sparrows with a normal 13 hour period of activity coincided the middle of this period with the middle of an artificial day whether it was of 1, 4 or 7 hours. The experimental evidence published was rather slight, but Aschoff (1965 a) has presented a much fuller case (nine species of birds) for the basic importance of the relation of the mid point of the activity time to the mid point of light time.

# CHAPTER II

# Field tests of theories of navigation
# by the sun's co-ordinates

We saw in the previous chapter that there is much field evidence that the sun is involved in bird navigation, probably in greater detail than as a simple compass reference. The several hypotheses of the ways in which the bird could utilize the sun's position for determining fixes are theoretically acceptable, though no decision has been reached as to which proposal is more plausible. We may now consider further evidence that *some* form of sun navigation is being used.

All the hypotheses postulate that the bird is comparing observed values with remembered values. This accords with the run of the homing experiments, which show that birds do *not* have innate knowledge of the ancestral home (and hence of its co-ordinates). They adopt as home the place in which they come to maturity. Once firmly attached to a place, generally after first breeding there, it is very difficult, if not impossible, to get them to home to any other place. The home co-ordinates must therefore be learned. If a complete form of sun navigation is used this implies that the birds must learn one (Matthews) or all (Pennycuick) of the characteristics of a family of sun arcs having in common the same symmetry with respect to noon in time and south in direction, but with the culmination point rising and falling with the seasons. It might well be that birds had inherited information on the general characteristics of the sun arc and its seasonal changes. It might also be that experience of the progressive changes with season would enable a form of prediction by the birds. But there was the possibility that comparison of observed values was made with values as last seen at home. Birds prevented from observing the sun arc for some time might then derive false information from its observations and comparisons. The fall in the culmination point with season (fig. 36) might not be distinguishable from a fall due to movement to a lower latitude.

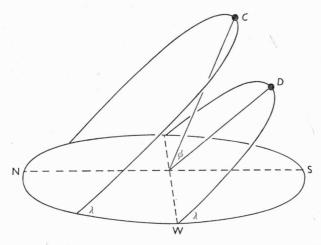

Fig. 36. Perspective diagram of the sun arc at summer solstice and at the equinoxes for 51° N. The highest point of the arc at $C$ has a greater altitude than at $D$ (by $\phi$) but the inclination of the arc ($\lambda$) remains the same. Compare with fig. 32.

During much of the year the change in the sun arcs is slow. Thus the culmination point changes only 10″ of arc per day at the solstices, but at the equinoxes the rate of change is 23′ of arc per day. Matthews (1953a) prevented Pigeons from having any view of the sun, sky or direct sunlight for six and nine days respectively over two autumnal equinoxes. The birds were then released south so that the net altitude (seasonal fall less latitude rise) of the culmination point was 1° 15′ and 2° 24′ lower than when they had last seen it. The results (fig. 37a) clearly suggest a difference between experimental and control birds. Thereafter it is a matter of statistical opinion. On a 't' test the experimental birds' scatter is not distinguished from random. On a '$X^2$' test they split into two groups, the majority with a southward tendency as would be expected if a net fall in the culmination point was interpreted as being due to transportation to the north.

Rawson & Rawson (1955) and Kramer (1955) repeated this 'waiting experiment' modified so that birds could see part of the sky during incarceration. They found no difference between controls and experimentals. Kramer (1957) returned to using light-proof conditions and his experimentals (no controls were used) had a predominant tendency towards home, in the north. But the general northward tendency shown by

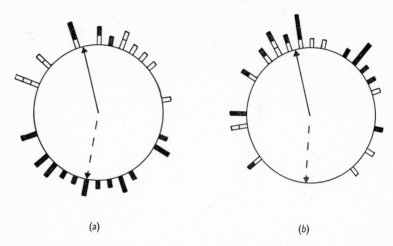

(a)                                             (b)

Fig. 37. Conflicting results obtained by preventing Pigeons, of related stock, from viewing the decline of the sun's noon altitude for some days at autumnal equinox and then releasing them south of home. In (a) most of the experimental birds appeared to vanish preferentially in the hypothetical false direction (dotted arrow), whereas the control birds headed homeward. In (b) no such distinction emerged. For discussion see text. (After Matthews, 1953a; Hoffmann, 1958.)

Wilhelmshaven Pigeons on release was by then realized, and simple sun-compass orientation would not, of course, be thrown out by such treatment. Hoffmann (1958) therefore sought to resolve the difficulty by operating from the loft near Cambridge originally used by Matthews. Not surprisingly Pigeons of Wilhelmshaven stock showed similar results to those obtained in Germany—Kramer (1959) reported that such birds even with no free-flying experience departed predominantly north-wards on release. Of more interest are the results of Pigeons derived from stock purchased from the same English fancier who had, six pigeon generations previous, supplied Matthews with his stock. They are shown in fig. 37b. The performance of the experimentals is *not* to be distinguished from that of the controls (though that is somewhat undistinguished) and we thus have a conflict of evidence. In view of what is now known of the variability in the orientation performances of Pigeons, and of the small numbers involved, it would not be wise to make a decision either way. Despite the rather optimistic title of Hoff-mann's paper, he used a different release point, 15 miles away, and used a different prior training technique, giving his birds

134

experience by releases only up to 14 miles in four compass directions. Matthews used the standard technique of training in one direction, in this case west, up to 80 miles. As we have seen (p. 82), such training may counteract directional tendencies.

Even if Pigeons are indeed not inconvenienced by being deprived of a sight of the sun for a few days, this does not imply that sun navigation is eliminated. There remain the possibilities of their applying a correction, innate or learned, for the seasonal changes. Again, determination of the inclination of the arc at release would indicate the true nature of the shift in latitude.

Another set of observations which may suggest that a form of astronavigation is used was afforded by the crater/palisade experiments started by Kramer (1959) and finished by Wallraff (1966b). We have seen (p. 89) that the results could be due to the elimination of landmark learning, though this is unlikely. Kramer (1959) concluded that the results indicated that the birds were relying on clues in the bottom 3° of the sky, that portion cut off by the palisade. However, Matthews (1959) suggested that the observed effects might well be due to interference with the practice in the use of a 'sextant', the bird's own movements producing rapid angular changes with reference to the available horizon, due to its closeness. This point was further stressed by Pennycuick (1960a).

Pigeons in the crater aviary were given a view of the distant horizon from a second storey. Their homing improved, 9 out of 125 returning, and the recoveries showed a distinct homeward bias similar to that of the open aviary birds. The palisade aviary offered the possibility of restoring the natural horizon in a graduated series of steps. A wide gap, 3·8 m. × 1·4 m., was opened in the upper part of the north side of the palisade. From various positions within the aviary a Pigeon could, in total, see over 140° of the horizon. Of 47 Pigeons reared in these conditions, none homed and the pattern of recoveries did not depart from a random scatter. More surprisingly, a similar gap opened in the south side of the palisade gave, with 118 Pigeons, equally poor results. Next, five gaps were opened in the upper part of the palisade, each about 1/15th of the perimeter, so that birds inside could see the complete horizon, though not simultaneously from any one point. In these circumstances 3 of 52

Pigeons homed, and although only eleven recoveries are available they showed a homeward tendency. This began to look as if as long as a Pigeon was able to set the whole of the distant visual horizon against the information of its own 'personal' horizon (derived from gravitational information), much of the disorientating effect of a palisade disappeared. However, the next step, to remove the whole of the upper part of the palisade and replace it with a series of framed glass windows, did not have the completely liberating effect expected. Two of 99 Pigeons did home, but the scatter of the recoveries was wide, with a homeward component (0·30) considerably less than that of the birds in the five-gap aviary or in the raised crater-aviary. It was, however, considerably greater than the near random scatters with the full palisade or with north or south openings. One could argue that the heavy frame of the windows in effect were providing a close horizon which might still cause difficulties of parallax. On the other hand, Wallraff has a point in suggesting that the glass might be causing an unspecified disruption of an unknown phenomenon.

The experimental approach was now reversed. The birds were allowed to see the whole of the distant horizon but their upward view was limited by a projecting opaque roof. From the perches they would only be able to see $4°$ above the horizon, momentarily in flight up to $14°$. Care was taken that the birds did not see the sun, a series of shutters being interposed at dawn and dusk. Thus in effect we have another 'waiting experiment'. Pigeons were kept in the roofed aviary from when they were four or five weeks old until release. Although they would have seen the sun from the breeding lofts, thereafter they were denied direct sight of it (though they could perhaps infer a good deal of its position from shadows). But as most were shut away in the spring and released when at least four months old, it could well be that the sun arc at release would be much the same as when they saw it before the solstice. Not one of the 174 Pigeons so treated homed. The recoveries are shown in fig. 38, and altogether do not make any marked homeward impression. However Wallraff reported that two different lighting regimes were used. In one the inner supplementary lighting was altered weekly to keep it in step with sunrise and sunset. In the other the lighting was kept fixed at 16 hours a day. The recoveries from the first treatment do show something more of a homeward

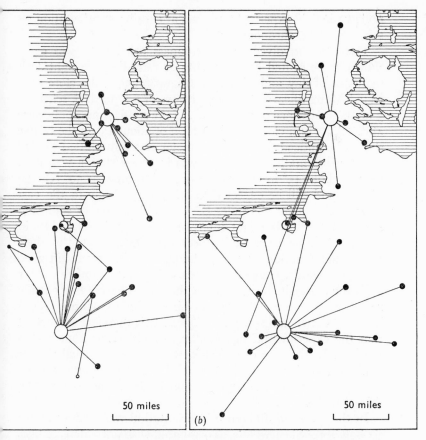

Fig. 38. Recoveries of Pigeons kept for months in a roofed aviary which prevented them seeing the sun but allowed an all round view of the horizon. A homeward tendency is discernible if the day length was maintained in phase with the normal (*a*), but not if a constant day length was employed (*b*). (From Wallraff, 1966*b*.)

trend, though not from the northern release point. These are odd results and Wallraff wisely is reserving detailed comment until further data are accumulated. Meanwhile it can be said that even if the suggestion that birds must be kept in touch with the natural horizon and the local time is confirmed, this does not rule out astronavigation—in which these are both essential ingredients.

In longitude determination we are concerned with the ability to know, and remember, the time of day at home; so that observations at release can be compared with those that would be expected at home at the same instant. We may therefore review the field experiments aimed at causing disorientation by shifting such a chronometer.

The clock involved in sun-compass orientation was early shown to be shifted by advancing or retarding the light/dark sequence of day and night. Matthews (1955 *a*) therefore subjected Manx Shearwaters to four days of a light/dark schedule 3 hours in advance of home time. If the sun-compass only were affected, an anti-clockwise deflection of about 45° in the course should be noticed. If the chronometer was fully shifted so that the sun information on release was interpreted as a westward shift of about 45° in longitude (some 2000 miles at 50° N), the birds would still be expected to orientate strongly to the east after displacement 265 miles E. A possible source of confusion was that the birds might interpret the experimental shift in light/dark sequence as having been brought about not by their chronometers suddenly being 3 hours slow, but by a shift 2000 miles eastwards. Generally the experimental chamber cannot be at the home site because of the difficulty of excluding outside noises, such as the calls of other birds, which Gwinner (1966) has now shown can serve as *Zeitgeber*. It is therefore important that the experimental surroundings should be as homelike as possible. In the case of the Shearwaters the nocturnal uproar of the colony is very noticeable, and a gramophone record of this was played to the experimental birds at the appropriate, shifted times.

In the event no difference in orientation was achieved with six batches of Shearwaters, as compared with untreated controls (155 birds in all). Weather conditions were far from perfect but the birds could not be kept longer in their boxes. Pigeons, however, could be incarcerated indefinitely and a group was now subjected (Matthews, 1955 *a*) to 3 hours advanced treatment for 10 days, before being released at an easterly point. No difference between experimentals and controls was apparent. This could have been due to the chronometer being more resistant to change than the sun-compass clock or to the misinterpretation of the imposed shift discussed above. An earlier experiment (Matthews, 1953 *a*) had shown that the chronometers of Pigeons could be thoroughly disturbed by treating them to 4 or 5 days with very irregular light/dark changes. This treatment was therefore now imposed before passing on to a regular series of days, time shifted by 3 hours. The results of three experiments in which Pigeons were brought on to 3 hours retarded days and then released at western points are shown in fig. 39. The westerly trend of the experimentals is clear. Another release of 3 hours advanced birds to the east gave, this time, an easterly trend. A simple compass shift would have directed the experimentals in fig. 39 clockwise 45° from the home direction, i.e. to between ESE and SSW.

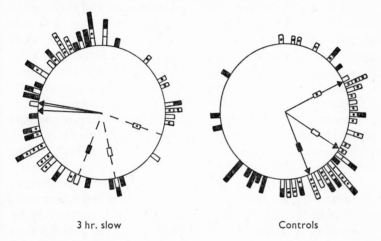

3 hr. slow                                        Controls

Fig. 39. Pigeons with their 'chronometers' 3 hours slow and released at three different places did not show a simple compass deviation of about 45° clockwise from the home direction (dotted lines). Rather, they showed a much greater westerly deviation as if the apparent forwardness of the local sun time was interpreted as due to their having been taken far to the east (with home directions as shown by the full lines). (After Matthews, 1955a.)

While only one experimental Pigeon returned on the day, against four controls, the difference in returns thereafter was not marked. Partly this may have been due to some of the experimentals not being influenced by the treatment and thus orientating well towards home. But even allowing for this, it is surprising that as much as one-third of the falsely-orientated experimentals returned on the second day (as against nearly half the homeward-orientated controls). Kramer (1957) stressed this point and also that, although the birds might well rapidly discard the artificial time schedule in favour of a natural time, the latter would be local not home time. One could still argue that the birds transported west with retarded clocks would only have the previous artificial schedule with which to make comparison and would then recognize that the time schedule in the place of incarceration *had* been, say, $2\frac{3}{4}$ hours slow, i.e. due to it being in the west, which would be corrected by flying east!

Schmidt-Koenig (1958, 1961 *b*) then embarked on a massive series of clock-shifting experiments in an attempt to clear the impasse. A total of 921 Pigeons was clock-shifted (with no difficulty, in 4 to 7 days) 6 hours early or 6 hours late or 12 hours out of phase. The scatters obtained were wide and their vector means were not so precisely in accordance with expectation as his trained birds in cages (p. 31), being respectively 72° anticlockwise, 93° clockwise and 168° clockwise. Nevertheless the imposed shifts were in accord with an effect on a simple sun-compass not on a longitude detecting mechanism. However the inference is not so conclusive as at first appears. The deviations of the experimental birds were not recorded in terms of compass points because the vector mean of the control birds diverged so widely from the home directions. The 'distance effect', now so amply confirmed by Schmidt-Koenig himself (p. 84) is the explanation here. More than half (474) of the experimental releases took place in the inner zone (5 to 14 miles) where good orientation towards home occurs. Since this is apparently by recognition of landmarks, and Graue (p. 85) has shown a sun-compass component in landmark-orientation, simple azimuth shifts would only be expected. Most of the other experimental releases were at distances (19 to 43 miles) which lie within the zone of disorientation with low homeward components, birds either showing random scatters or orientation in northward 'nonsense' or other odd directions. Only 6 % of the releases took place at distances (58 to 103 miles) where the controls would be expected to show a strong homeward component.

Of course, once outside the landmark zone, the experimental birds might well have been expected to show a tendency east or West after their apparent enormous excursions in longitude, 4000 miles, imposed by 6-hour shifts. After 12-hour shifts confusion should have resulted since they would apparently be on the opposite side of the world. Pennycuick (1961) suggested that such displacements are indeed so gross and 'impossible' that they would be ignored by the Pigeons. They might then fall back on such 'nonsense' tendencies as they had, and these being based on a sun-compass would be deviated predictably from the controls. The way the bulk of Schmidt-Koenig's data are presented, with reference to the (unspecified) vector means of the controls released on the same day, make it impossible to re-interpret them in terms of compass directions. A decision cannot therefore be reached as to whether the longitude 'information' was accepted and reacted upon, caused general confusion, or was ignored. The data from five release points, of the seventeen used, where a connection with compass directions can be derived, suggest the latter two possibilities as the more likely.

Recently Walcott & Michener (1967) have embarked upon experiments using time shifts of 'reasonable' dimensions, 10 minutes to 2 hours. Though few in number their results have added importance in that their Pigeons were followed home by tracking the radio-transmitters they bore. A complication was introduced by flying each bird several times from one point before releasing it at another after clock-shifting treatment. Thus the bird had a conflict of choice between flying in the compass training direction and correcting the longitudinal shift indicated by the clock shift. The experimenters deliberately imposed small time shifts which should not alter the sun compass very much but which would nevertheless represent substantial shifts in longitude. Releases were at distances of 35 to 65 miles, i.e. probably outside the zone of known landmarks. Two birds shifted only 10 minutes initially adhered very close to the learned compass bearing but homed very slowly. Four birds shifted 15 to 20 minutes showed a much greater deviation (33° to 82°) from the learned bearing than would be expected on a simple sun-compass hypothesis. Three birds shifted two hours showed still greater deviations (91 to 121°) even though the expected deviation (30°) was also greater. In these cases the initial track lay between the training and the expected longitude-correction direction, indeed favouring the latter. The authors tentatively concluded that shorter time shifts were more disturbing to the Pigeons' navigation than the longer shifts. This view is strengthened by further unpublished results (Wallcott, personal communication). Thus, while these are only first indications from this sophisticated technique, they do strongly indicate that Pigeons use their clocks for navigational purposes other than, and in addition to, the sun-compass. Another point on which these tracking experiments accord with those of Matthews is that, despite the initial perturbation of their course the birds homed quite rapidly after the 2-hour shifts.

# CHAPTER 12

# The anatomical and physiological
# limitations of the avian eye

Both the hypotheses of bi-coordinate sun navigation require the bird to detect and measure the sun's movement along its arc. This requirement has often been thought to be quite impossible by opponents of the hypotheses. The impression was given that they had thrown the whole question out of court by stating that the bird is required to detect movement slower than that of the hour hand of a watch. However, response to movements slower than this have been demonstrated by investigators of the optomotor reaction (leg movements in relation to movements of a striped cylinder) in insects. We may cite the results of McCann & MacGinitie (1965) with *Musca*, and Mittelstaedt (1964) with *Stagomantis*—who reports responses to movements as slow as 2° 20′/hour, less than one-sixth that of an hour hand. Horridge (1966*a*, *c*, *d*) in a most elegant series of experiments has shown that the eyes of the crab *Carcinus* can detect and follow the movements of a small light down to velocities as slow as 6–8°/hour. This was in an otherwise dark room so that other objects could not serve as stabilizing reference points. Moreover he demonstrated that the crab's eyes could detect and follow the sun's actual movement along its arc, beginning only 10 seconds after first seeing the sun.

Even with relatively crude techniques Meyer (1964) has now found a similar high degree of movement detection ability in Pigeons. Again he used birds with successful homing experience, and presented them with a 'sun' projected on a curved homogeneous surface. Their own movement was restricted. By commencing with the 'sun' moving at 110°/hour and then, in subsequent test series, reducing the speed to 55°/hour, 25°/hour and finally 15°/hour, he found that the birds could clearly distinguish between non-movement and movement even at the latter speed (that of the sun along its arc). He did not test at lower speeds, leaving a possibility that some other clue, such as the noise of the 'sun'-moving mechanism, was reacted to, rather than movement *per se*. It is also not clear to what extent the apparatus provided frames of reference (which would greatly facilitate movement detection). Another indication of the movement detection capabilities of the pigeon eye is provided by flicker-fusion studies such as those of Dodt & Wirth (1953) and of Granit (1959). It was found that Pigeons could resolve flashing rates as high as 150 per second, as compared with 60 per second for humans.

The avian eye is thus undoubtedly better than the human eye in detecting movement. For the latter Leibowitz (1955) claimed, under practically ideal conditions, 15°/hour *with* a point of reference. Boyce (1965) concludes that the ultimate human limit, *without* any external reference, is between 60 and 120°/hour. The reason is undoubtedly the very different structure of the eyes, as discussed by Pumphrey (1961). There are a number of features in the avian eye that probably enhance its powers of movement detection. First, nearly all the retinal surface lies in the image plane, so that all distant objects are sharply focused. Secondly, there is much less difference between the concentration of cones in the *fovea* and in the rest of the retina than there is in the human eye. The net result of these two features is that the bird can with one glance take in a picture which a man can only accumulate by laboriously scanning the whole field, piece by piece. Thirdly, Pumphrey (1948) has suggested that the characteristically steep profile of the central *fovea* can, by imposing a momentary distortion, enhance the appearance of movement. Fourthly, Menner (1938) suggested that the *pecten*, primarily a nutritive organ, enhances movement detection by casting a foliated shadow on parts of the retina. He showed the foliations to be particularly well developed in species dependent on the capture of moving prey, and constructed a model of the apparatus that effectively demonstrated that movements were indeed more visible. Crozier & Wolfe (1943 *a*, *b*) using a flicker-fusion technique confirmed Menner's postulates. Probably the shadow increases the contrast as the observed object moves across the field.

Lastly, Maturana and Frenk (1963), using micro-electrodes to register the electrical output of individual ganglion cells in the pigeon retina, have demonstrated the existence of at least two types responding selectively to movement. One type, the directional movement detectors, form about 30 % of the accessible cell population. They have small receptive fields, $\frac{1}{2}$ to 1° in diameter, and give an optimal, or exclusive, response to the movement of the edge of a small object moving in one direction, but not in the reverse. In general the size of response (number of spikes and frequency) depends on the speed of movement, as well as contrast factors. If a spot of light is directed on the receptive field, then cut off and moved before being directed on it again, a response is obtained if the movement was in one direction, not if it was in the opposite. A second type of ganglion cell, forming only 5 % of the accessible population, responds maximally, when the head is in normal orientation to gravity, to a horizontal edge moved vertically up or down. It does not respond to small objects or to on/off lights. Maturana & Frenk (1965) describe the centrifugal fibres in the pigeon retina which, because of their great number and their even distribution, appear to constitute a system of localized control of retinal functions. Undoubtedly further interesting discoveries will follow future probings on the cellular level.

It is a strong point in favour of the hypotheses of astronavigation that they rely on a sense organ that is particularly well developed in birds, and do not have to invoke unknown sense organs. In passing to a further consideration of the potentialities of the avian eye, it is interesting to note that a tradition has developed among pigeon fanciers that good homers can be detected by their 'eye sign' (Bishop, 1954). However, the 'sign' is provided by shadowy marks on the iris, adjacent to the pupil, whose nature or relation to function are not explained.

There seems no doubt that we can allow that the birds can detect the movement of the sun. How about the requirement, under one hypothesis, for the extrapolation of movement? Meyer (personal communication) is working on a development of his apparatus that will test this aspect. Meanwhile we can point out that extrapolation of the path of moving objects is essential in birds feeding on moving prey, such as a plunging Gannet. Conversely, birds flying fast in close proximity to objects must be able to judge their own future track with accuracy. The 'good eye' of an expert in ball games or of the crack shot depends on an ability for extrapolation. In the case of the sun, too, the problem is much simplified by the fact that it is an object moving at a constant speed in a limited number of ways and without any complications of parallax.

That the basic requirements for extrapolation exist in arthropod eyes at least is suggested by Horridge's (1966*a*, *b*) findings. If the stimulus was switched off, moved slightly and then switched on after an interval, his crabs could remember the original position to within 0.5° after 2 minutes, to within 1° after 15 minutes. Locusts could remember the stimulus position to within 0·1° for many seconds.

Relevant to this problem of extrapolation is the proven ability of some animals to estimate parts of the sun path which they have not seen. Lindauer (1957) showed that bees which continued to perform their direction-indicating dances through the night indicated correctly the angle between the daytime food source and the azimuth of the sun at the particular time of night. Lindauer (1959) after allowing his bees to observe the sun only in the afternoons found that they were nevertheless able to use it for navigation immediately they were released one morning. Similarly, Fischer (1961) has shown that lizards compass-trained at different times of day to an artificial sun can construct the appropriate azimuth course at other times. Even more remarkable are the conclusions of New & New (1962) concerning the dances of bees in equatorial regions when the sun is close to the zenith, and they could only have the vaguest impression of the sun azimuth. Yet in these conditions they still give directed dances. Apparently they are systematically turned in a direction 'which could not have been learnt by observing sun movements, and which may not correspond with the changing sun azimuth occurring at the time, but which nevertheless takes into account dance directions (or sun azimuths) memorized for other times of day, and between successive memorized positions turns the dances through an angle proportional to time. In fact they appear to have an innate mechanism that can divide angles by time'.

The time-interval measurement aspects of extrapolation and rate-of-change appreciation are considered in the next chapter, here we may note that they seem to present no obstacle. Turning to the necessary discrimination of small differences in stimulus position, the basic requirement is a fine 'grain' in the retina. However, in passing we may note that compound eyes of locusts (Burtt & Catton, 1962) and crabs (Horridge, 1966c) have been shown capable of discriminating stimuli only a fraction of the inter-ommatidial angle apart. Barlow (1965) has stressed that this is only possible when the stimulus is moving. We may agree with Horridge that 'the acuity of the input side remains astounding'. In the case of diurnal birds the retinae are almost pure cone in composition and the visual cells attain their densest known packing—1 million per mm²—at the bottom of the *fovea* in the largest hawks. This is four to five times the density in the human *fovea*. To quote Polyak (1957) 'in order to remain with accepted concepts we may consider the possibility that in the case of keen sighted species, such as the (Golden) Eagle, light of short wave lengths can still be comfortably accommodated within cones measuring not much more in thickness than 1/3 micron (corresponding to a visual angle of 4 seconds of arc) and is preponderantly or exclusively utilized'. Even more important, a high density of cones is retained over the rest of the retina in birds, whereas in humans it has fallen to 4% of the foveal value a few millimetres away. The bird is thus capable of a high degree of resolution over the whole retina, without having to fixate on the objects to be discriminated.

The density of the nuclei in the ganglion cell layer, which is much thicker than in our own, indicates that there is practically a 1:1 correspondence with the cone cells. Since acuity depends on each nerve fibre responding to the stimulation of the smallest possible retinal area, a lack of sensory summation (which would lead to increased light sensitivity, and occurs in the rods of nocturnal eyes) is another contribution to the efficiency of the avian eye for our present purpose. Walls & Judd (1933) have suggested that the yellow oil droplets present in the cones have the function of suppressing the chromatic fringes which would hinder acute discrimination. They are also thought to be useful in cutting down glare (Walls, 1963).

The retinal structure thus gives promise of a high visual acuity, perhaps exceeding that of the human by three or four times. But this would not be realized if the image projected on the retina were inadequate. The requirement is for a big image covering as many visual cells as possible, and this means a large eye, large absolutely, rather than relatively, for cone size does not vary greatly. In the birds evolution has certainly favoured the emergence of huge eyes, as large as the head can carry. An Ostrich has eyes five times the size of a human's, a large hawk's are about the same as ours. In such birds superiority in retinal structure should indeed be reflected in superior visual acuity. The eyes of the smaller birds are inevitably smaller, but relatively the Starling's, for instance, are fifteen times as massive as a man's.

Another requirement to make full use of a fine retinal structure is a dioptric apparatus with a large aperture—the ratio of pupil diameter to focal length (Lord, 1956). In birds this is achieved by the flattened shape of the eye-ball coupled with a small pupil, which also helps acuity by cutting down spherical aberration at the edge of the lens. Since the muscles of the iris are, uniquely, striated, the size of pupil is under 'voluntary' control. The irises of the two eyes react independently to stimuli, though some light leakage to the opposite orbit may cause a delayed reaction (Levine, 1955). The reaction time is much faster than in humans (Bishop & Stark, 1965). So here again we have qualitative and quantitative differences in the visual systems of bird and mammal.

A man's eye is capable of resolving two points about 30″ apart, a performance rather better than could be explained on the basis of the size and numbers of the foveal cones. The improvement is thought to be achieved by the activity of the 'associational' cells of the inner layers of the retina. (Spence (1934) gives a figure of 26″ for the Chimpanzee.) There appear to be no experimental data on acuity in the larger birds' eyes. Donner (1951) investigated the situation in a number of passerines by requiring them to distinguish between a grid of black and white lines and a grey object of the same brightness. The Reed Bunting had the poorest discrimination (between 3′ 50″ and 3′ 07″) followed by Yellowhammer (3′ 07″ and 2′ 38″) and Robin (2′ 38″ and 1′ 55″). Song Thrush, Fieldfare and Chaffinch gave the best results (1′ 20″ and 40″). These figures were in agreement with the 'morphological' acuity indicated by the foveal structure. On this latter basis Oehme (1962) investigating Swift, Starling and Blackbird estimated a *minimum separabile* of 1′ 10″. In the case of Pigeons there is a conflicting range of values obtained by various investigators, ranging from 7′ 48″ (Chard, 1939) through 3′ 12″ to 2′ 42″ (Hamilton & Goldstein, 1933) and 1′ 32″ (Walls, 1963) to 23″ (Grundlach, 1933). These investigators did not have the advantage of the more sophisticated operant conditioning techniques, and indeed the elegant 'threshold tracking' method developed by Blough (1956) has yet to be applied to the visual acuity problem. Tansley (1965) states that a homing Pigeon will resolve two lines whose retinal images are less than $1\ \mu$ apart, whereas the human figure is $1\cdot89\ \mu$. There is little doubt that an acuity similar to man's (and over the whole of the retina, not just in the *fovea*) is a fair assumption to make for our purposes, bearing in mind a morphological *minimum separabile* of only $10''$ in some species.

Of course the ability to discriminate two simultaneously presented points or lines is not the same thing as detecting small changes in large angles, one value being remembered. But the excellence of the bird's eye has been stressed since as a sense organ it seems fully capable of the types of performance required in astronavigation. Thus *if* a difference in the sun's noon-altitude or arc angle of $1'$ could be appreciated, navigation to within a mile or two would be possible. But as we have seen (p. 83) the 'distance-effect' means that we do not have to look for navigational ability below about 50 miles. Thus we are far from approaching the threshold in this matter. Meyer (1964) set up an operant-conditioning experiment whereby Pigeons (experienced homers) were required to discriminate an artificial sun moving at $15°$/hour presented at an altitude of $45°$ from one presented on other occasions at $15°$, $10°$, $5°$, $1°$, $30'$ and $15'$ higher. Successful discrimination was established down to a $30'$ difference in altitude, corresponding to about 35 miles. Next Meyer tested the Pigeons' ability to discriminate a difference in arc angle, any cue from height being eliminated by the experimental procedure. He concluded that his birds discriminated down to and including $1°$ in difference, but not to $30'$. At $45°$ latitude this would represent ability to detect a displacement of about 40 miles. Thus the experimental results seem to be closely in accord with the findings in the field. Further experiments of this nature would clearly be of the greatest value for there are apparent weaknesses in the technique as reported (p. 143).

Pennycuick (1960a), in considering the likelihood that accurate enough measurements could be made to accommodate his version of the sun-navigation hypothesis, pointed out that a human subject in the dark can set a luminous line to the vertical with an accuracy better than 1° (Noble, 1949). If a bird could do likewise an accuracy of 69 miles could be obtained. This ability has not been tested to threshold in birds, though Zeigler & Schmerler (1965) and Mello (1965) found that Pigeons, unlike some animals, readily discriminate between rectangles orientated in different directions, having no difficulty with discriminations involving obliques. Meyer's results also suggest that the threshold will be found to be low, certainly lower than the 4° found for bees (in flight) by Wehner & Lindauer (1966). Mello's curious finding, that, if only one eye is trained to an oblique (say 45°), the other eye when tested responds maximally to the mirror image (135°), only serves to emphasize the unusual nature of the visual system. It is not due, as Cumming *et al.* (1965) suggested, to an inter-ocular projection through the inter-orbital septum into the back of the untrained eye. With regards the accuracy required in measuring rate of change of altitude, Pennycuick presented calculations suggesting that an accuracy of about $1\frac{1}{2}$ seconds of arc per minute of time would be required to give navigational accuracy of about 14 miles. He had proposed that as a bird circles the release point the sun's image traces a line on the retina. Then as the bird enters another circle, another line is traced. The sensory problem is then to determine the distance and time interval separating the lines. Barlow (1965) points out that the human eye can detect a line 1″ in width, about 30 times *less* than the *minimum separabile* of two points. Taking this for the bird to be 10″, the ability to locate a line should be of the order of $\frac{1}{3}$″, with a probable error in the difference between two lines of $\frac{2}{3}$″. To get the necessary accuracy the bird would be required to take two observations at a precisely measured time interval (see p. 155) of 30 seconds. Pennycuick considered the performance required of the bird to be 'physically possible and quite reasonable on the information available'. Even if his *minimum separabile* figure is optimistic by, say, a factor of three, the navigational accuracy he is working to is conservative by a similar amount, so the net requirement would be similar.

As at least limited navigation appears possible by the stars and moon we should consider finally vision at night. The requirements for this conflict sharply with those for acute day vision. One reason why the birds' daytime eyes are better than humans' is that they have evolved straight through for daylight vision, instead, as in our own case, of going through a nocturnal stage and then reorganizing the eye back to daytime requirements, with all the inevitable compromises of efficiency. A night eye must be sensitive to small quantities of light. A big pupil is needed; this necessitates a large lens, often nearly spherical, if problems of refraction at the periphery are to be overcome. A nocturnal retina is dominated by rods, whose connections show a great deal of summation, such that several thousand may be linked to each optic nerve fibre. Rods take a long time to become dark adapted, but the increase in sensitivity may be more than a million-fold. Cones dark-adapt in a few minutes but sensitivity only increases about a thousand times. Where there is a mixed retina there is a characteristic 'step' in the process of adaptation. The typical diurnal bird's eye with small pupil and lens, and a nearly pure cone retina, will be relatively insensitive at night. This has been confirmed by Blough (1956), Adler & Dalland (1959) and Adler (1963) in the Pigeon, Starling and American Robin. The first named is rather more sensitive when dark-adapted than the other two, but even so its sensitivity is well below that of a human.

Specifically in the context of navigation, it is likely that diurnal birds will see fewer stars than we can ourselves. This is not necessarily a disadvantage, for it is easier for us to pick out the main patterns of the constellations when the carpet of lesser stars is somewhat dimmed by haze. From the practical, experimental, point of view, however, it means that clouds that will obscure stars from us will do so from birds. Birds which habitually fly at night as well as by day can be expected to have more rods than diurnal birds. Lockie (1952) showed that this was so when comparing the Manx Shearwater's retina with that of a House Sparrow, but the difference was not great, and became negligible when compared with a diurnal sea bird of the same family, the Fulmar Petrel. The Shearwater does show a rather greater degree of summation of visual cells to ganglion nuclei. Certainly anyone who has been blundered into by a Shearwater in the dark will have appreciated the poorness of their night vision. Matthews (1955b) has suggested that they locate their nesting holes by smell if their mates are absent and thus not providing welcoming shouts. The retina of the Mallard which actively forages at night would repay investigation.

We have seen (p. 43) that the same individual Mallard can orientate both by the sun and under a starry sky. It is possible that the lack of indications of homeward orientation at the distances of displacement so far investigated (158 miles) may reflect a lower acuity due to the compromise necessary to provide adequate sensitivity at night, leading to an inability to appreciate small changes in the sun's co-ordinates.

# CHAPTER 13

# Motion, time and memory

Another feature of the avian eye structure which may have a relevance to navigation is the shape of the central *area* of high cone density (in which the *fovea* is situated). In many birds of open spaces the *area* is elongated along the horizontal equator of the eye. This is so in the flamingos, shearwaters, gulls, waders, geese and cormorants (but not in the pigeons). Pumphrey (1948) suggested that 'such a design appears suited to fixation of the horizon and to effecting a preferential increase in sensitivity to vertical movements of objects in relation to the horizon'. Duijm (1959) showed that the apparent deviation from the horizontal in some species is corrected by the position in which the head is normally held. Certainly there is a temptation to regard the ribbon-like *area* as the equivalent of an aircraft's 'artificial horizon', expecially as Duijm (1951) has also demonstrated that the horizontal semi-circular canal of the inner ear is horizontal during the attitude of general alertness, not that of rest. Lowenstein (1950) notes that the *cristae* of the (phylogenetically more recent) horizontal canal differ in structure from those of the other two and responses are to rotations in its own plane within narrow limits—again differing from the wide range of receptivity of the vertical canals. Pennycuick (1960 a) was concerned with the possibilities of the *area* helping to establish the vertical with reference to *visual* features, and here the discovery of horizontal edge movement detectors in the pigeon retina (p. 144) may well be relevant. He pointed out that with some landscapes ('electric braes') the information as to what is horizontal could be misleading, possibly giving rise to the poor results in certain localities reported by Kramer (p. 110). However, undoubtedly a central problem is the extent to which birds can determine the vertical when in flight, *independently* of vision, i.e. can they disentangle the effects of acceleration, rotation and other changes in posture from the downward effect of gravity?

Many authorities are emphatic that such an achievement is not possible, generally citing the difficulties the human aviator can get into when flying blind, if he did not have his 'instruments'. This is a curious argument when a very small part of the human species has been piloting and navigating airplanes for less than seventy years, barely time for a third generation to emerge, while birds have been depending for their existence on flight for some 140 million years. A bird moreover *is* the airplane, not like the human pilot, shut up in a box and deprived of any sensory contact with his environment other than that provided by gravity. Birds, too, have been more modest in the speeds they attain and hence in the forces they impose in turns. Because a human pilot loses track of which way up he is in a cloud this is emphatically no argument that a bird likewise is disorientated. But it must also be admitted that practical evidence on what a bird *is* capable of sensing *in flight* is sadly lacking. It may well be that the advent of micro-miniature telemetry will begin to give us some answers. It is certainly a fascinating and virtually untouched field. Even on the ground, most of the sophisticated experimentation has been on vertebrates other than birds (see e.g. Lowenstein, 1950).

In normal flight a bird's head is quite remarkably stable and does not reflect the violent oscillations undergone by the body in the wing section. This impression can be confirmed by studying cine films, particularly those of long-necked birds like flamingos, and by careful measurements, frame by frame. Brown (1953) published pictures of Pigeons in vigorous flight just after take-off in enclosed surroundings, showing the eye centre moving up or down in a ratio of as little as 1:330 of the forward movement. The Kestrel hovering with its head motionless in a strong, gusty wind is a most remarkable sight. Another is afforded by wild geese 'whiffling' in to land at a safe roost. At first glance it appears that the birds execute a full sideways roll over on their backs in flight. But high speed film reveals that it is the body only that rolls through 180°, the head proceeding forward on a level keel the normal way up (Wallin, 1962).

Of course in these manoeuvres the bird has the use of its eyes to help it maintain its head position, even though there does not seem to be an absolute need for fixing rigidly on some distant object—the Kestrel, for instance, can be seen moving its head from side to side as it scans the ground below. One is glad to note that there has been a proper reluctance to experiment with permanently blinded birds. But it is surprising that more has not been done with opaque hoods that a bird can easily scratch off when it has landed. Hochbaum (1955) reports such experiments on eighteen species in the plains of Manitoba, involving 249 individual flights. When thrown up blindfold 95% quickly adopted a normal flying position and flew for varying periods, two fifths for more than a minute and some going out of sight. Ducks often flew in an apparently entirely normal manner, as did many of the smaller birds, blackbirds and sparrows. Among the latter many adopted a hovering flight. Although flight action appeared normal, the birds often flew in circles and were drifted downwind. Of those which successfully flew only 20% crash-landed. Most spiralled gently down at a shallow angle or parachuted with hovering wings and out-thrust legs. Landings were not orientated with regard the wind, indeed, as was to be expected, the birds showed no appreciation of wind direction except momentarily in gusts. This confirmed some few experiments by Beecher (1952) who flew blindfolded birds on the end of a thin string.

Birds, then, can maintain stable, level flight in the absence of visual stimuli. What is not clear is how far this is due to the inherent stability of the flight-structure of the bird and how far it is due to rapid correction of displacement by some form of 'feed-back' mechanism; whether, indeed, the sensing mechanism is confined to the semi-circular canals and utriculi of the inner ears. It has long been known that a blind-folded bird held in the hand keeps its head level, and makes compensatory movements with wings and legs no matter how its position is shifted. Thus a tilt to the right extends the right leg and wing and flexes and lifts the left leg and wing. Trendelenburg (1906) concluded that, as the wing responses of chickens occurred regardless of head position, the sensors detecting the tilts imposed must be in the body. He satisfied himself that this was so by sectioning certain thoracic dorsal roots, whereupon the response failed. Paton (1928) came to the conclusion that, with ducks, visceral sensors were also concerned. However, Mittelstaedt (1964), in a series of elegant mathematical arguments, showed that the observed results could be explained equally well as due to the interaction of two sets of receptors, the one measuring the deviation of the head with respect to gravity, the other (in the neck) the deviation of the head from the body. The interpretation agreed even better with the facts if (another example of Mittelstaedt's general bi-component hypothesis) each of the two deviations is represented by two messages, one proportional to the sine, the other to the cosine of the angle. Yet when Mittelstaedt later tested the responses of Pigeons after total labyrinthine extirpation, they proved qualitatively and quantitatively the same as in the normal birds. So, after all, these wing and leg reflexes, at least, are mainly or exclusively elicited by gravity receptors in the body.

A precise estimation of speed, and, since velocity is the time taken to travel a given distance, of short time intervals, is implicit in extrapolation of the sun's movement, just as it is of successful flight in flocks, in the presence of obstacles and when landing. It should be noted, however, as Gregory (1966) has pointed out, that in some retinae movement is coded therein (p. 144) or immediately behind (in the visual projection areas of the brain) and therefore velocity can be perceived without involving an estimate of time in each case. Like a car's speedo-meter, a clock is needed for calibrating the instrument, there-after it measures velocity without recourse to a clock.

Endogenous neural rhythms have received considerable atten-tion (e.g. Bullock, 1962) and underlie locomotor rhythms from the relatively slow walking of the chick, about 2·2 cycles/second (Bangert, 1960) to the fast wing beats of humming birds, up to 50 cycles/second (Stolpe & Zimmer, 1939). Other movements likely to show temporal stereotypy are those involved in court-ship signals. Thus Dane et al. (1959) found the 'Simple Head Throw' movement of the Goldeneye's repertoire had an average length of 1·29 seconds and a standard deviation of only 0·08 seconds in sixty-six examples. The inability of Stein (1951) to get finches to adopt a feeding rhythm of less than four minutes pales beside the performances extracted from Pigeons by Ferster & Skinner (e.g. 1957) in which birds can be taught to play duets on the keys of their operant conditioning boxes.

A more natural type of duetting occurs in the songs of many birds and involves timing of a very precise nature. Thorpe (1963 b) found that in a pair of the Black-headed Gonolek, which call out of sight of each other, the bird producing the second note in the duet did so at an average interval of 0·425 seconds after the first bird's initial note, with a standard deviation (seven consecutive duets) of only 0·0049 seconds. A pair of Chubb's Cisticola, duetting in sight of each other, had an interval of 0·396 seconds with the remarkably low standard deviations of 0·0029 seconds (eight times better than a human could achieve in similar circumstances). Another requirement for precise measurement of the interval between the production of a sound, and, in this case, its return, is implicit in the successful echo-location ability shown in complete darkness by Oil Birds (Griffin, 1953) and Swiftlets (Medway, 1959). Such performances are, of course, surpassed by those of the bats (see e.g. Griffin, 1958; Grinnell, 1963; Pye, 1963) which may well be responding to echoes of sounds emitted only 0·001 seconds earlier.

As an illustration of visual discrimination of temporal order we may cite Robinson (1967). Humans were shown in flashes triangles and squares, and required to say in which order they came. Under dichoptic (as opposed to binocular) conditions three subjects perceived the presentation order correctly when the interval between flashes was the lowest tried, 0·005 seconds.

The detection and measurement of short time intervals by animals is thus wholly within the bounds of possibility. Next we must enquire whether there are biological clocks known that will serve as longitude chronometers. These must be able to tell the bird on release its home time to within a very few minutes. The 'distance effect' suggests that Pigeons must be able to detect longitudinal displacements equivalent to to 40 to 60 miles at latitudes 40 to 55° N. This represents about 1° in longitude or 4 minutes in time. Biological clocks which maintain their period in uniform conditions of light and temperature with this degree of constancy have now definitely been proven to exist. Thus DeCoursey (1961) reports on eighteen Flying Squirrels, *Glaucomys*, held in constant darkness for periods of 10 to 81 days and whose activity cycles had standard deviations of ±2 to ±9 minutes. An example is shown in fig. 40. Many others can be found in the literature referred to on p. 30. The length of the free-running cycles under constant conditions can be changed e.g. by changing the light intensity (Hoffmann, 1960a). But this does not imply a modification of the underlying cyclic mechanism. Rather the threshold for the emergence of activity has been raised or lowered (Aschoff, 1967). Similarly, variation in the period length under a given set of constant conditions reflects not only the clock's precision but also variation in the processes linking the clock information to the onset of activity. Thus the clock's own precision would be even better than the $\pm 2$ minutes referred to above.

Biological rhythms of this nature are now recognized to be

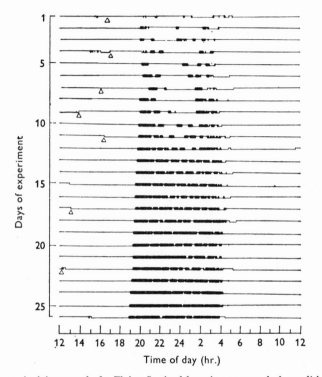

Fig. 40. Activity record of a Flying Squirrel kept in constant dark conditions for 26 days. A circadian rhythm of 23 hours 58 minutes ($\pm 4$ minutes) was maintained, and drifted very slowly out of phase with the external time. The triangles indicate brief, irregular maintenance interruptions. (From DeCoursey, 1961.)

pretty well universal, down to unicellular animals. Indeed in higher animals it is now thought that the controlling mechanisms are at the cellular level also. They are now called *circadian* rhythms since it became apparent that most of those studied were not of exactly 24 hour length but about (*circa*) a day (*dies*) long. The cycle lengths vary not only between species, but also within species. Thus DeCoursey's squirrels had periods between

22 hours 58 minutes and 24 hours 21 minutes, the most common class being in the range 23:50 to 23:59. This feature is of great theoretical importance since it indicates that the rhythms are indeed biological, based on endogenous oscillating processes and not kept in time by reference to some all-pervading geophysical controlling stimulus (e.g. magnetism or cosmic rays), as postulated by Brown (e.g. 1965). Additionally, Brown's experimental results have been shatteringly re-analysed by Enright (1965). However, Wever (1967) does claim that human circadian rhythms are influenced by weak magnetic fields.

No matter how *constant* the clock is, if its period differs from 24 hours it will drift out of phase with sun time (just as the star sphere 'drifts' westward with the seasons, p. 43). So clocks of exactly 24 hour periodicity will be at a premium in species capable of bi-coordinate navigation. There seems no good reason why such periodicity should not be selected for in view of its survival value. None of the long-distance navigators have yet been tested to ascertain the constancy and periodicity of their clocks.

Stein (1951) having demonstrated that six species of passerines could be trained to take food at a particular hour, found that two Siskins adhered to the training time through 8 days of constant light. The increase in activity before the feeding time was not sufficiently precise a measure to determine whether drift occurred. Hoffmann (1954) used a more precise indicator of time-keeping by training two Starlings to take food in a particular direction with reference to the sun. They were then kept for 28 days in conditions of constant light and temperature. Tested at irregular intervals they maintained the correct direction of choice (and hence a correct time) until the end of the experiment. In this case the clocks did not drift out of phase, and the experiments of Gwinner (1966) also involved two Starlings which had free-running activity rhythms very close to 24 hours. On the other hand, Hoffmann (1960b) experimented with two other Starlings whose activity period was about $23\frac{1}{2}$ hours. After 12 days in constant light their direction choice had deviated through about 60° anti-clockwise. This is in the direction anticipated, but to a lesser degree, from clocks drifting half an hour a day.

Little has been done to extend Stein's work to ascertain the accuracy of time-keeping under normal light/dark conditions.

Adler (1963) tied the bird's reactions more closely to the food-getting by requiring it to peck a key which delivered food only at certain periods of the day. As no punishment was given for pecking out of turn the errors of three Starlings were invariably anticipatory ones. For this reason their average value, 15 to 20 minutes, can only be considered an outside figure for accuracy. Meyer (1966 a) carried out further tests on homing Pigeons with prior successful homing experience. Although this latter does not necessarily demonstrate bi-coordinate navigational ability, such screening is better than expecting every individual of a species to show the ability. The birds were on a 12/12 hours light/dark sequence with masking 'white' noise. Food was available in response to key tapping at a fixed time after the house lights came on. The intervals tested were half, one and two hours and it was found that 'with a composite for the three time intervals and for the data from the two subjects, it was estimated that the errors in time judgements for 45° N would give rise to a displacement of the bird between 50 and 60 miles on the east-west axis'. The need remains for further tests in which 'dawn' does not come at the same time as during the training period, to check how important this may be as a reference point. It would not appear essential since Stein (1951) trained seven birds in constant light *ab initio*. However, this does not prove that a change in the light/dark sequence would not shift the clock.

Indeed, in view of the ease with which the clock controlling the sun-compass reaction can be shifted (p. 31) the question is frequently raised whether a bird has a chronometer sufficiently rigid, i.e. uninfluenced by outside events, to measure displacement in longitude. One answer might be the existence of two time-keepers, one, easily malleable, the clock for sun-compass orientation, the other, more rigid, the chronometer for longitude measurement. The existence of more than one time-keeper in an animal is now readily acceptable. Indeed the trend in theoretical thinking on the subject is towards a multiplicity of time-keepers, perhaps ultimately at the cellular level, which may or may not phase in or be coupled with one another, or as Pittendrigh (1960) puts it, 'the organism comprises a population of quasi-autonomous oscillatory systems'. A very elegant experimental demonstration of two time-keeping mechanisms involved in the control of activity in the cockroach *Periplaneta* has been given by Harker (1964). Renner (1959) provides evidence of more than one clock governing the food-seeking of bees *Apis*. Trained at a fixed time and then translocated through 49° of longitude ($\equiv 3\frac{1}{4}$ hours) the bees showed two peaks of activity. One corresponded to the time before the shift and was relatively inflexible, the other drifted fairly quickly (within 3 days) to the new time according to external conditions. A long series of experiments have convinced Gwinner (in Aschoff, 1967) that the timing of migratory restlessness in birds can best be explained on a model with *two* coupled oscillators. Aschoff (1965 $b$) reports a human subject in which the circadian rhythms for activity and calcium excretion differed widely from those of body temperature, potassium and water excretion.

The relative inflexibility of Matthews' Shearwaters and Pigeons to a change in light regime (p. 138) may therefore reflect a qualitative difference in time-keeping mechanisms. Certainly if there are rigid as well as constant time-keeping mechanisms we would expect to find them in birds capable of full navigation. But is it necessary to insist on a more-or-less permanent rigidity adduced by critics of bi-coordinate sun-navigation? The requirement is that, on release, the bird should detect the displacement it has undergone. This it should be able to do if its chronometer is sufficiently rigid to last through the period of transportation (experimental or natural) in which it does not have access to astronomical information. Even if it were not able to leave the release area at once, and its chronometer started to get in phase with local time, the direction and perhaps distance of home would have already been determined. A useful experiment here would be to transport birds and then hold them in conditions such that the local day/night *Zeitgeber* could influence them but they could not see the sun. If the chronometer were flexible then the ability to detect the displacement in longitude should decline (unless the bird remembered the difference between the first actual and expected dawn/dusk).

We must not, however, lose sight of the possibility of extremely rigid clocks maintaining home time over say half a year. Serventy (1963) has demonstrated a remarkable rigidity in the egg-laying of the Short-tailed Shearwater in S.E. Australia. On one island 85 % of the birds lay their single egg within

3 days of the peak dates 25–26 November, which historical evidence indicates as having been the same for over a century. Individual females laid on exactly the same day in four successive years. Yet in the non-breeding period these birds migrate right up to the Aleutian Islands and back, through a wide range of day lengths and time zones. Even more remarkable, Marshall & Serventy (1959) held Shearwaters captive for months under varying light regimes and found that their gonads developed at just the same rate as did those of the birds which undertook the great migration. Their conclusion, that there is an extremely precise, persistent timing mechanism requiring synchronisation no more than once a year, seems as unassailable as the annual 'clock' itself.

The last subject we need to touch on, having a bearing on the hypotheses of sun navigation, is that of memory. Thorpe (1963 a) quotes many examples of the excellence of memory, both auditory and visual, in birds of many species. Hardy (1951) lists some apparently well-authenticated cases of Pigeons returning to their home loft after absences of up to eight years. Skinner (1950) taught a Pigeon to peck at a particular feature on an aerial photograph, and then found perfect retention after four years. Indeed the achievements of operant conditioning with Pigeons (e.g. Ferster & Skinner, 1957) show again and again the ability of the birds to memorise an enormous variety of features of the environment, and the correct response thereto. Now Meyer (1964, 1966 a) has used the technique to prove the ability of Pigeons to remember the angular features of a 'sun' arc and of time intervals.

It is thus no strain on our credulity to require birds to remember the characteristics of the sun arc prior to displacement, for the few days involved in the usual experimental situation. The requirement is rather tougher when we are considering migrational homing after an interval of half a year. However, the inclination of the sun arc remains constant for a given latitude, and although the culmination altitude varies it will have a similar value on the birds' return in the spring as it did on their departure in the autumn. It is often the long-distance migrants that are precise in their arrival and departure. Thus the Great Shearwater with its remarkable navigational problem (p. 1) departs, according to Rowan (1952), in the second week of April from its isolated breeding grounds and returns there at the beginning of September, dates ten weeks on either side of the summer solstice. The return is much earlier than 'necessary' since the eggs are not laid until November—when they appear in a rush, apparently rivalling that of the Short-tailed Shearwater. Corti (1931) and Geyr (1943) give other examples of seasonal symmetry in migration.

Memory of longitude, or the time-correlated aspects of the sun arc characteristics, imposes more difficulties. The implication would be that the breeding-grounds time is maintained on a rigid 'chronometer' through many months. Of course much migration is essentially north/south, the birds remaining in the same time band, and no great difficulty is involved. However, many birds, such as the Eurasian ducks and geese, undertake migrations in which the main displacement is east/west. It would seem unreasonable on the present evidence to require such birds to maintain a time out of schedule with the surroundings for such long periods. The proven existence of an annual clock of high constancy (p. 160) does not necessaily mean that it measures shorter intervals, of a day, with similar constancy. But this very difficulty perhaps gives us the clue as to the part played by homing ability in the natural order of things, uninfluenced by the machinations of pigeon-fanciers or experimental zoologists.

We have seen that young migrants are apparently innately equipped with enough information to carry out bearing-and-distance navigation to the wintering grounds, possibly flying a series of courses rather than one. The return to the breeding grounds is accomplished by flying the reciprocal courses. Older birds possess the full navigational ability and are able to correct for involuntary displacements such as are produced by beam winds encountered over the sea or at night. This possibility for correction of migratory courses may well be the main natural function of bi-coordinate navigation, whether it is effected by a return to the point before displacement occurred, or by a vector change in the migration direction. After all, land birds do not forage, nor are likely to be blown, the sort of distances from the nest at which bi-coordinate navigation is now thought to become effective—fifty miles or so. Sea-birds probably do make longer journeys, and such species have been shown experimentally to be the best navigators. Even these should probably be thought of as starting out in a chosen direction and perhaps integrating voluntary changes of course to maintain an awareness of the direction in which to return home. Only when the bird 'loses track' of its surroundings through a storm, say, would the ability to detect displacement from its last 'fix' be at a premium. Nevertheless involuntary displacements would occur sufficiently frequently to give the development of such an ability a selective advantage.

On this reasoning bi-coordinate navigation is to be thought of as a flexible ability with the function of enabling the bird to return to where it should be at any particular season. It is not a device tied, with the bird, rigidly to one point in time and space, the breeding area.

In Pigeons the ability to detect displacement is inborn, or at least matures within a few months. However, inexperienced birds rarely home, presumably due to the lack of urge to return or to their being easily led to join other pigeons over whose lofts they fly. The displacement experiments still leave the possibility that young migrants *may* detect a sideways shift, but not act to correct it owing to the over-riding influence of passing conspecifics. Even if this were so, it carries no implication of their having inherited information concerning the *locality* of a place to which they have never been.

The ways in which the various forms of navigational ability, from fixed angle orientation through to bi-coordinate navigation, can have become fitted together in the course of phylogeny (or, for that matter, ontogeny) remain matters for fascinating speculation.

# SCIENTIFIC NAMES OF SPECIES MENTIONED

| | |
|---|---|
| Albatross, Laysan | *Diomedea immutabilis* |
| | |
| Blackbird, Yellow-headed | *Xanthocephalus xanthocephalus* |
| Red-winged | *Agelaius phoeniceus* |
| Blackcap | *Sylvia atricapilla* |
| Bluethroat | *Cyanosylvia svecica* |
| Bobolink | *Dolichonyx oryzivorus* |
| Brambling | *Fringilla montifringilla* |
| Bunting, Indigo | *Passerina cyanea* |
| Ortolan | *Emberiza hortulana* |
| Reed | *Emberiza schoeniclus* |
| | |
| Chaffinch | *Fringilla coelebs* |
| Cisticola, Chubb's | *Cisticola chubbi* |
| Coot | *Fulica atra* |
| Cowbird | *Molothrus ater* |
| Crossbill | *Loxia curvirostra* |
| Parrot | *Loxia pityopsittacus* |
| Crow, Hooded | *Corvus cornix* |
| Prairie | *Corvus brachyrhynchos* |
| Cuckoo | *Cuculus canorus* |
| Black-billed | *Coccyzus erythrophthalmus* |
| Bronze | *Chalcites lucidus* |
| Yellow-billed | *Coccyzus americanus* |
| Curlew, Bristle-thighed | *Numenius tahitiensis* |
| | |
| Dove, Mourning | *Zenaidura macroura* |
| Duck, Wood | *Aix sponsa* |
| Dunnock | *Prunella modularis* |
| Eagle, Bald | *Haliæetus leucocephalus* |
| Golden | *Aquila chrysætos* |
| | |
| Fieldfare | *Turdus pilaris* |
| Flycatcher, Pied | *Muscicapa hypoleuca* |
| Collared | *Muscicapa albicollis* |
| Frigate Bird | *Fregeta magnificens* |
| | |
| Gannet | *Sula bassana* |
| Goldeneye | *Bucephala clangula* |
| Goldfinch | *Carduelis carduelis* |

| | |
|---|---|
| Gonolek, Black-headed | *Laniarius erythrogaster* |
| Goose, Barnacle | *Branta leucopsis* |
| Canada | *Branta canadensis* |
| Goshawk | *Accipter gentilis* |
| Greenfinch | *Chloris chloris* |
| Grosbeak, Rose-breasted | *Pheucticus ludovicianus* |
| Scarlet | *Carpodacus erythrinus* |
| Gull, Black-headed | *Larus ridibundus* |
| Common | *Larus canus* |
| Greater Black-backed | *Larus marinus* |
| Herring | *Larus argentatus* |
| Lesser Black-backed | *Larus fuscus* |
| | |
| Hawfinch | *Coccothraustes coccothraustes* |
| Hawk, Sparrow | *Accipter nisus* |
| | |
| Kestrel | *Falco tinnunculus* |
| | |
| Lark, Western Meadow | *Sturnella neglecta* |
| Linnet | *Carduelis cannabina* |
| | |
| Mallard | *Anas platyrhynchos* |
| Martin, Purple | *Progne subis* |
| House | *Delichon urbica* |
| | |
| Oil Bird | *Steatornis caripensis* |
| Ostrich | *Struthio camelus* |
| Ousel, Ring | *Turdus torquatus* |
| | |
| Penguin, Adelie | *Pygoscelis adeliae* |
| Petrel, Fulmar | *Fulmarus glacialis* |
| Leach's | *Oceanodroma leucorrhea* |
| Storm | *Hydrobates pelagicus* |
| Pigeon, Homing | *Columba livia* |
| Rock | *Columba livia* |
| Pintail | *Anas acuta* |
| Pipit, Meadow | *Anthus pratensis* |
| Tree | *Anthus trivialis* |
| Plover, Golden | *Charadrius dominicus* |
| | |
| Redstart | *Phoenicuras phoenicurus* |
| Robin | *Erithacus rubcuela* |
| American | *Turdus migratorius* |
| | |
| Shearwater, Great | *Procellaria gravis* |
| Manx | *Procellaria puffinus* |
| Short-tailed | *Procellaria tenuirostris* |
| Sooty | *Procellaria grisea* |

| | |
|---|---|
| Shelduck | *Tadorna tadorna* |
| Shrike, Lesser Grey | *Lanius minor* |
| Red-backed | *Lanius collurio* |
| Siskin | *Carduelis spinus* |
| Skua, Great | *Catharacta skua* |
| Sparrow, Eastern Tree | *Spizella arborea* |
| Fox | *Passerella iliaca* |
| Golden Crowned | *Zonotrichia atricapilla* |
| House | *Passer domesticus* |
| Tree | *Passer montanus* |
| White Crowned | *Zonotrichia leucophrys* |
| Starling | *Sturnus vulgaris* |
| Stork, White | *Ciconia ciconia* |
| Swallow | *Hirundo rustica* |
| Bank | *Riparia riparia* |
| Cliff | *Petrochelidon pyrrhonota* |
| Swift | *Apus apus* |
| Alpine | *Apus melba* |
| Swiftlet | *Collocalia maxima* |
| | |
| Teal, Blue-winged | *Anas discors* |
| Green-winged | *Anas crecca* |
| Tern, Arctic | *Sterna macrura* |
| Common | *Sterna hirundo* |
| Noddy | *Anous stolidus* |
| Sooty | *Sterna fuscata* |
| Thrush, Grey-cheeked | *Hylocichla minima* |
| Swainson's | *Hylocichla ustulata* |
| Thrush-Nightingale | *Luscinia luscinia* |
| | |
| Veery | *Hylocichla fuscescens* |
| | |
| Wagtail, White | *Motacilla alba* |
| Warbler, Barred | *Sylvia nisoria* |
| Garden | *Sylvia borin* |
| Grasshopper | *Locustella naevia* |
| Wood | *Phylloscopus sibilatrix* |
| Whinchat | *Saxicola rubetra* |
| Whitethroat | *Sylvia communis* |
| Lesser | *Sylvia curruca* |
| Wryneck | *Jynx torquilla* |
| | |
| Yellowhammer | *Emberiza citrinella* |

# REFERENCES

The following list contains only works referred to in the preceding text and, like it, aims at being comprehensive but not exhaustive. It is hoped that it includes all accounts of substantial experimental work on bird navigation, and papers setting out original theoretical contributions. In the interests of space a substantial number of more trivial references included in the First Edition are here omitted.

Preliminary announcements, summaries and translations are excluded in favour of the main publication concerned. Papers which are reiterations of previous publications are represented only by the most comprehensive of the series. No attempt has been made to include unoriginal commentaries, which are legion.

A number of good textbooks on spherical trigonometry and astronavigation are available. For a general introduction E. W. Anderson (1967) *The Principles of Navigation*, London, is recommended, and not only for its dedication 'To the Manx Shearwater that, taken from his nest off the coast of Wales and carried three thousand miles across the Atlantic to Boston, was back in twelve days'.

AAGAARD, J. S. & WOLFSON, A. (1962). Transister equipment for continuous recording of oriented migratory behaviour in birds. *IRE Trans. Bio-Med. Electron.* **9**, 204–8.

ADAMS, D. W. H. (1962). Radar observations of bird migration in Cyprus. *Ibis*, **104**, 134–46.

ADLER, H. E. (1963). Psychophysical limits of celestial navigation hypotheses. *Ergbn. Biol.* **26**, 235–52. (also *Anim. Behav.* **11**, 566–77.)

ADLER, H. E. & ADLER, B. P. (1966). Computer simulation of pigeon homing. *Abstracts XIV Int. Orn. Cong. Oxford*, 25.

ADLER, H. E. & DALLAND, J. I. (1959). Spectral thresholds in the starling (*Sturnus vulgaris*). *J. Comp. & Physiol. Psychology*, **52**, 438–45.

AGRON, S. L. (1963). Evolution of bird navigation and the earth's axial precession. *Evolution*, **16**, 524–7.

ARDREY, R. (1966). *Territorial Imperative*. London.

ARNOULD-TAYLOR, W. E. & MALEWSKI, A. N. (1955). The factor of topography in bird homing experiments. *Ecology*, **36**, 641–6.

ASCHOFF, J. (1965a). The phase-angle difference in Circadian periodicity. In *Circadian Clocks* (ed. Aschoff). Amsterdam. 262–76.

ASCHOFF, J. (1965b). Circadian rhythms in Man. *Science*, **148**, 1427–32.

ASCHOFF, J. (1967). Circadian rhythms in birds. *Proc. XIV Int. Orn. Cong. Oxford*, 81–105.

BALL, S. C. (1952). Fall bird migration on the Gaspé Peninsula. *Peabody Mus. Nat. Hist. Yale Univ. Bull.* **7**, 1–211.

BANGERT, H. (1960). Untersuchungen zur Koordination der Kopf- und Beinbewegungen beim Haushuhn. *Z. Tierpsychol.* **17**, 143–64.

Barlow, H. B. (1965). Visual resolution and the diffraction limit. *Science*, **149**, 553-5.

Barlow, H. B., Kohn, H. I. & Walsh, E. G. (1947). Visual sensations aroused by magnetic fields. *Amer. J. Physiol.* **148**, 372-5.

Barlow, J. S. (1964). Inertial navigation as a basis for animal navigation. *J. Theor. Biol.* **6**, 76-117.

Barnwell, F. H. & Brown, F. A. (1964). Responses of planarians and snails. In *Biological effects of magnetic fields* (ed. M. J. Barnothy). New York. p. 263-78.

Barrett, W. (1883). Notes on the alleged luminosity of the magnetic field. *Phil. Mag. J. Sci.* **15**, 270-5.

Barthel, R. & Creutz, G. (1959). Verfrachtung von Heckenbraullen (*Prunella modularis*). *Vogelwarte*, **20**, 38-9.

Bateson, P. P. G. & Nisbet, I. C. T. (1961). Autumn migration in Greece. *Ibis*, **103**a, 503-16.

Batschelet, E. (1965). Statistical methods for the analysis of problems in animal orientation and certain biological rhythms. *A.I.B.S. Monograph*. Washington D.C.

Becker, G. (1963). Ruheeinstellung nach Himmelsrichtung, ein Magnet-Orientierung bei Termiten. *Naturwiss.* **50**, 455.

Becker, G. & Speck, U. (1964). Untersuchungen über de Magnetfeld-Orientierung von Dipteren. *Z. vergl. Physiol.* **49**, 301-40.

Becker, R. O. (1963). The biological effects of magnetic fields—a survey. *Med. Electron. Biol. Engng.* **1**, 293-303.

Beecher, W. J. (1951). A possible navigation sense in the ear of birds. *Amer. Midl. Nat.* **46**, 367-83.

Beecher, W. J. (1952). The role of vision in the alighting of birds. *Science*, **115**, 607-8.

Bellrose, F. C. (1958a). The orientation of displaced waterfowl in migration. *Wilson Bull.* **70**, 20-40.

Bellrose, F. C. (1958b). Celestial orientation in wild Mallards. *Bird Band.* **29**, 75-90.

Bellrose, F. C. (1963). Orientation behaviour of four species of waterfowl. *Auk*, **80**, 257-89.

Bellrose, F. C. (1967a). Radar in orientation research. *Proc. XIV Int. Orn. Cong. Oxford*, 281-309.

Bellrose, F. C. (1967b). Orientation in waterfowl migration. *Proc. Ann. Biol. Coll. Oregan State Univ.* **27**, 73-99.

Bellrose, F. C. & Graber, R. R. (1963). A radar study of the flight direction of nocturnal migrants. *Proc. XIII Int. Orn. Cong. Ithaca*, 362-89.

Benjamins, C. E. (1926). Y a-t-il une relation entre l'organe paratympanique de Vitali et le vol des oiseaux? *Arch. néerl. Physiol.* **11**, 215-22.

Bergman, G. (1964). Zur Frage der Abtriftskompensation des Vogelzuges. *Orn. Fenn.* **41**, 106-10.

Bergman, G. & Donner, K. O. (1964). An analysis of the spring migration in Southern Finland. *Acta Zool. Fenn.* **105**, 1-59.

Bernis, F. (1966). *Migracion en Aves*. Madrid.

Billings, S. M. (1968). Homing in Leach's Petrel. *Auk* **85**, 36-43.

Birukow, G. (1956). Lichtkompassorientierung beim Wasserläufer *Velia*

*currens* F. am tage und zur Nachtzeit. I. Herbstund Winterversuche. *Z.*
*Tierpsychol.* **13**, 463–84.

BIRUKOW, G. & BUSCH, E. (1957). Lichtkompassorientierung beim Wasserläufer *Velia currens* F. (Heteroptera) am Tage und zur Hachtzeit. II.
Orientierungsrhythmik in Verschiedenen Lichtbedingungen. *Z. Tierpsychol.* **14**, 184–203.

BISHOP, L. G. & STARK, S. (1965). Pupillary response of the Screech Owl,
*Otus asio. Science,* **148**, 1750.

BISHOP, S. W. E. (1954). *The Secret of Eye-Sign.* London.

BLOESCH, M. (1956). Algerische Störche für den Störchansiedlungsversuch
der Vogelwarte Sempach. *Orn. Beob.* **53**, 97–104.

BLOESCH, M. (1960). Zweiter Bericht über den Einsatz algerischer Störche
für den Störchansiedlungversuch in der Schweiz. *Orn. Beob.* **57**, 214–23.

BLOUGH, D. S. (1956). Dark adaptation in the pigeon. *J. Comp. physiol.
Psychol.* **49**, 425–30.

BOCHENSKI, Z., DYLEWSKA, M., GIESZCZYKIEWICZ, J. & SYCH, L. (1960).
Homing experiments on birds. XI. Experiments with Swallows *Hirundo
rustica* L. concerning the influence of earth magnetism and partial eclipse
of the sun on their orientation. *Zesz. Nauk. W.J. Zoologica,* **5**, 125–30.
(In Polish, English summary.)

BOUCHNER, M. & SEDIVY, J. (1959). Versuche über das Orientierungsvermögen des Haus- und Feldsperlings (*Passer domesticus* und *P. montanus*).
*Sylvia,* **16**, 185–202.

BOYCE, P. R. (1965). The visual perception of movement in the absence of
an external frame of reference. *Optica Acta,* **12**, 47–54.

BOYD, H. (1964). Barnacle Geese caught in Dumfriesshire in February 1963.
*Wildfowl Trust Ann. Rep.* **15**, 75–6.

BRAEMER, W. (1959). Versuche zu der im Richtungsfinden der Fische
erhaltenen Zeitschätzung. *Verh. dtsch. Zool. Ges. Zool. Anz 23 Supplement-
band,* 276–88.

BRAEMER, W. (1960). A critical review of the sun-azimuth hypothesis. *Cold
Spring Harbour Symp.* **25**, 413–27.

BRAEMER, W. & SCHWASSMANN, H. O. (1963). Von Rhythmus der Sonnenorientierung am Äquator (bei Fischen). *Ergebn. Biol.* **26**, 182–201.

BROWN, C. E. (1939). Homing pigeons exposed to radio frequency waves.
*Sci. Amer.* **160**, 45.

BROWN, F. A. (1965). A unified theory for biological rhythms: rhythmic
duplicity and the genesis of 'circa periodisms'. in *Circadian Clocks.* (Ed. J.
Aschoff) Amsterdam. 231–61.

BROWN, R. H. J. (1953). The flight of birds. II. Wing function in relation
to flight speed. *J. Exp. Biol.* **30**, 90–103.

BUB, H. (1962). Heimfindeversuche mit Haussperlingen in Nordwest-
deutschland. *Falke,* **9**, 164–71.

BULLOCK, T. H. (1962). Integration and rhythmicity in neural systems.
*Amer. Zool.* **2**, 97–104.

BÜNNING, E. (1964). *The Physiological Clock.* Berlin.

BURTT, E. T. & CATTON, W. T. (1962). A diffraction theory of insect vision.
I. An experimental investigation of visual acuity and image formation
in the compound eye of three species of insects. *Proc. Roy. Soc.* B **157**, 53–82.

BUSNEL, R. G., GIBAN, J., GRAMET, P. & PASQUINELLY, F. (1956). Absence d'action des ondes du radar sur la direction de vol de certains oiseaux. *Comptes Rendus des Séances de la Société de Biologie,* **150,** 18.

CASAMAJOR, J. (1927). Le mystérieux 'sens de l'espace'. *Rev. sci.* **65,** 554–65.

CASAMAJOR, J. (1930). *La Nature,* **2834,** 504–6. Quoted by Thompson (1947).

CASEMENT, M. B. (1966). Migration across the Mediterranean observed by radar. *Ibis,* **108,** 461–91.

CATHELIN, F. (1935). Rôle primordial des grands courants aériens électro-magnétique de profondeur dans la genèse des migrations des oiseaux. *L'Oiseau,* **5,** 284–91.

CHARD, R. D. (1939). Visual acuity in the pigeon. *J. exp. Psychol.* **24,** 588–608.

CLAPAREDE, E. (1903). La faculté d'orientation lointaine. *Arch. Psychol. Geneva,* **2,** 133–80.

CLARK, C. W. (1933). Night-flying homers of the signal corps. An experiment that resulted in a new race of homing pigeons. *Nat. Hist. (New York),* **33,** 409–18.

CLAUSEN, M., KOLLER, G. & KUHNEN, H. (1958). Unterliegt die Heim-kehrgeschwindigkeit der Brieftauben hormonalen Einflüssen. *Experientìa,* **14,** 386–8.

COCHRAN, W. W. & GRABER, R. R. (1958). Attraction of nocturnal migrants by lights on a television tower. *Wilson Bull.* **70,** 378–80.

COCHRAN, W. W., MONTGOMERY, G. G. & GRABER, R. R. (1967). Migratory flights of *Hylocichla* Thrushes in spring: a telemetry study. *Living Bird,* **6,** 213–25.

CORTI, U. A. (1931). Zeitsymmetrie im Vogelzug. *Orn. Beob.* **28,** 4.

CREUTZ, G. (1941). Ergebnisse de Verfrachtung von Grünfinken (*Chloris c. chloris*). *Vogelring,* **13,** 33. (*Vogelzug,* **13,** 89).

CREUTZ, G. (1949a). Verfrachtungen mit Kohl- und Blaumeisen (*Parus m. major* L. und *Parus c. caeruleus* L.). *Vogelwarte,* **15,** 78–93.

CREUTZ, G. (1949b). Zur Lebensweise des Feldsperlings. Untersuchungen zur Brutbiologie des Feldsperlings. *Zool. Jahrb. Abt. System.* **78,** 133–72.

CREUTZ, G. (1957a). Verfrachtungs-Versuche mit einem Grünfinken (*C. chloris*). *Vogelwarte,* **19,** 58–9.

CREUTZ, G. (1957b). Freilassung von Bergfinken (*Fringilla montifringilla*) nach der Zugzeit. *Vogelwarte,* **19,** 59–60.

CREUTZ, G. (1961). Nochmals: Freilassung von Bergfinken (*Fringilla montifringilla*) nach der Zugzeit. *Vogelwarte,* **21,** 53–4.

CROZIER, W. J. & WOLFE, E. (1943a). Theory and measurement of visual mechanisms. X. Modifications of the flicker response contour, and the significance of the avian pecten. *J. Gen. Physiol.* **27,** 287–313.

CROZIER, W. J. & WOLFE, E. (1943b). Flicker response contours for the sparrow, and the theory of the avian pecten. *J. Gen. Physiol.* **27,** 315–24.

CUMMING, W. W., SIEGEL, I. M. & JOHNSON, W. (1965). Mirror-image reversal in pigeons. *Science,* **149,** 1518–19.

CYON, E. V. (1900). Ohrlabyrinth, Raumsinn und Orientierung. *Pflügers Archiv.* **76,** 211–302.

DAANJE, A. (1936). Haben die Vögel einen Sinn für den Erdmagnetismus, wie Deklination, Inklination und Intensität? *Ardea,* **25,** 107–11.

DAANJE, A. (1941). Heimfindeversuche und Erdmagnetismus. *Vogelzug*, **12**, 15–17.

DANE, B., WALCOTT, C. & DRURY, W. H. (1959). The form and duration of the display actions of the Goldeneye (*Bucephala clangula*). *Behaviour*, **14**, 265–81.

DARWIN, C. (1873). Origin of certain instincts. *Nature, Lond.* **7**, 417–18.

DEBENEDETTI, E. T. (1962). Orientational response of some Amphipods under artificial light. *Boll. Ist. Mus. Zool. Univ. Torino*, **6**, 1–8.

DECOURSEY, P. J. (1962). Effect of light on the Circadian activity rhythm of the flying squirrel, *Glaucomys volans*. *Z. vergl. Physiol.* **44**, 331–54.

DESBOUVRIE, J. (1889). Quoted *Zoologist*, **47**, 397.

DIJKGRAAF, S. (1946). Over het orientatie probleem bij vogels. *Proc. Kon. Nederl. Akad. Watensch.* **49**, 690.

DIJKGRAAF, S. (1953). Unempfindlichkeit für langwelliges Licht beim Staren (*Sturnus vulgaris* L.). *Experientia*, **9**, 222.

DINNENDAHL, L. & KRAMER, G. (1950). Heimkehrleistungen italienischer und deutscher Reisetauben. *Vogelwarte*, **15**, 237–42.

DIRCKSEN, R. (1932). Die Biologie des Austernfischers, der Brandseeschwalbe und der Kustenseeschwalbe. *J. Orn.* **80**, 427–521.

DODT, E. & WIRTH, A. (1953). Differentiation between rods and cones by flicker electroretinography in pigeon and guinea pig. *Acta Physiol. Scand.* **30**, 80–9.

DOLNIK, V. R. & SHUMAKOV, M. E. (1967). Testing the navigational abilities of birds. *Bionica* Moscow. 500–07 (in Russian).

DONNER, K. O. (1951). The visual acuity of some passerine birds. *Acta Zool. Fennica*, **66**, 1–40.

DORST, J. (1962). *The Migrations of Birds*. Kingswood, Surrey.

DROST, R. (1934). Über Ergebnisse bei Verfrachtungen von Helgoländer Zugvögeln. *Proc. VIII Int. Orn. Cong. Oxford*, 620–8.

DROST, R. (1938). Über den Einflus von Verfrachtungen zur Herbstzugzeit auf den Sperber *Accipter nisus* (L.). *Proc. IX Cong. Orn. Int. Rouen*, 502–21.

DROST, R. (1949). Zugvögel perzipieren Ultrakurzwellen. *Vogelwarte*, **15**, 57–9.

DROST, R. (1955). Wo verbleiben im Binnenland frei aufgezogne Nordsee-Silbermöwen. *Vogelwarte*, **18**, 85–93.

DROST, R. (1958). Über die Ansiedlung von jung ins Binnenland verfrachteten Silbermöwen (*Larus argentatus*). *Vogelwarte*, **19**, 169–73.

DRURY, W. H. (1959). Orientation of Gannets. *Bird Band.* **30**, 118–19.

DRURY, W. H. & KEITH, J. A. (1962). Radar studies of songbird migration in coastal New England. *Ibis*, **104**, 449–89.

DRURY, W. H. & NISBET, I. C. T. (1964). Radar studies of orientation of songbird migrants in southeastern New England. *Bird Band.* **35**, 69–119.

DUCHÂTEL, M. (1901). Cited Claparède (1903).

DUIJM, M. (1951). On the headpostures of birds and its relation to some anatomical features. *Proc. Kon. Nederl. Akad. Wetensch.* **54**, 3–24.

DUIJM, K. (1959). On the position of a ribbon-like central area in the eyes of some birds. *Arch. néerl. Zool.* **13**, *Suppl. 1*, 128–45.

DYER, M. I. (1967). Photo-electric cell technique for analysing radar film. *J. Wildlife Manag.* **31**, 484–91.

EASTWOOD, E. (1967). *Radar Ornithology*. London.

EASTWOOD, E., ISTED, G. A. & RIDER, G. C. (1960). Radar 'Ring Angels' and the roosting movements of Starlings. *Nature, Lond.* **186**, 112–14.

EASTWOOD, E. & RIDER, G. C. (1964). The influence of radio waves upon birds. *Brit. Birds,* **57**, 445–58.

EASTWOOD, E. & RIDER, G. C. (1965). Some radar measurements of the altitude of bird flight. *Brit. Birds,* **58**, 393–425.

EASTWOOD, E. & RIDER, G. C. (1966). Grouping of nocturnal migrants. *Nature, Lond.* **211**, 1143–6.

EDWARDS, J. & HOUGHTON, E. W. (1959). Radar echoing area polar diagrams of birds. *Nature, Lond.* **184**, 1059.

ELDAROV, A. L. & KHOLODOV, YU. A. (1964). The effect of a constant magnetic field on the motor activity of birds. *Zhurnal Obschchei Biologii,* **25**, 224–9 (in Russian).

EMEIS, D. (1959). Untersuchungen zur Lichtkompassorientierung des Wasserläufers *Velia currens* F. *Z. Tierpsychol.* **16**, 129–54.

EMLEN, J. T. & PENNEY, R. L. (1964). Distance navigation in the Adelie Penguin. *Ibis,* **106**, 417–31.

EMLEN, S. T. (1967a). Orientation of *Zugunruhe* in the Rosebreasted Grosbeak, *Pheucticus ludovicianus. Condor,* **69**, 203–5.

EMLEN, S. T. (1967b). Migratory orientation in the Indigo Bunting, *Passerina cyanea.* Part I: Evidence for use of celestial cues. Part II. Mechanism of celestial orientation. *Auk,* **84**, 309–42, 463–89.

EMLEN, S. T. & EMLEN, J. T. (1966). A technique for recording migratory orientation of captive birds. *Auk,* **83**, 361–7.

ENRIGHT, J. T. (1961). Lunar orientation of *Orchestoidea corniculata. Biol. Bull.* **120**, 148–56.

ENRIGHT, J. T. (1965). The search for rhythmicity in biological time-series. *J. Theoret. Biol.* **8**, 426–68.

EVANS, G. (1795). *A Discourse on the Emigration of British Birds.* London.

EVANS, P. R. (1966a). Migration and orientation of passerine night migrants in northeast England. *J. Zool. Lond.* **150**, 319–69.

EVANS, P. R. (1966b). An approach to the analysis of visible migration and a comparison with radar observations. *Ardea,* **54**, 14–44.

EVANS, P. R. (1968). Re-orientation of small passerine night migrants after displacement by the wind. *Brit. Birds.* (in press).

EXNER, S. (1893). Negative Versuchsergebnisse über das Orientierungsvermögen der Brieftauben. *SitzBer. Akad. Wiss. Wien.* **102**, 318–31.

EXNER, S. (1905). Über das Orientierungsvermögen der Brieftauben. *SitzBer. Akad. Wiss. Wien,* **114**, 763–90.

FARNER, D. S. (1967). The control of avian reproductive cycles. *Proc. XIV Int. Orn. Cong. Oxford,* 107–33.

FATIO, V. (1905). Le sens de l'orientation, *Rev. sci. Paris,* **22**, 282.

FERGUSON, D. E., LANDRETH, H. F. & TURNIPSEED, M. R. (1965). Astronomical orientation of the Southern Cricket Frog, *Acris gryllus. Copeia,* **1**, 58–66.

FERSTER, C. B. & SKINNER, B. F. (1957). *Schedules of Reinforcement.* New York.

FISCHER, K. (1961). Untersuchungen zur Sonnenkompassorientierung und

Laufaktivität von Smaragdeidechsen (*Lacerta viridis* Laur.). *Z. Tierpsychol.* **18**, 450–70.

FRISCH, K. V., LINDAUER, M. & SCHMEIDLER, F. (1960). Wie erkennt die Biene den Sonnenstand bei geschlossener Wokendecke? *Naturwiss. Rundschau.* **13**, 169–72.

FROMME, H. G. (1961). Untersuchungen über das Orientierungsvermögen nächtliche ziehender Kleinvögel (*Erithacus rubecula, Sylvia communis*). *Z. Tierpsychol.* **18**, 205–20.

GEBEL, R. K. H., DEVOL, L. & WYLIE, L. R. (1960). Astronomical observations by means of highly sensitive electronic light intensification. *WADC. Tech. Note*, 59–404.

GEHRING, W. (1963). Radar- und Feldbeobachtungen über den Verlauf des Vogelzuges im Schweizerischen Mittelland: der Tagzug im Herbst (1957–1961). *Orn. Beob.* **60**, 35–68.

GEHRING, W. (1967a). Radarbeobachtungen über den Vogelzug am Col de Bretolet in den Walliser Alpen. *Orn. Beob.* **64**, 133–145.

GEHRING, W. (1967b). Analyse der Radarechos von Vögeln und Insekten. *Orn. Beob.* **64**, 145–51.

GERDES, K. (1962). Richtungstendengen vom Bratplatz verfrachteter Lachmowen (*Larus ridibundus* L.) unter Auischluss visueller Gelande-und Himmelsmarken. *Zeit Wiss Zool.* 166, 350–410.

GEYR VON SCHWEPPENBURG, H. (1922). Zur Theorie des Vogelzuges. *J. Orn.* **70**, 361–85.

GEYR VON SCHWEPPENBURG, H. (1943). Zur Zeitsymmetrie im Vogelzuge. *Vogelzug*, **14**, 112–13.

GEYR VON SCHWEPPENBURG, H. (1963). Zur Terminologie und Theorie der Leitlinie. *J. Orn.* **104**, 191–204.

GIBAULT, G. (1928). L'orientation du pigeon voyageur et les phénomènes magnétiques, électriques et météorologiques. *La Nature*, **2788**, 17–19.

GIBAULT, J. (1930). Recherches sur l'orientation du pigeon voyageur. *C.R. Congr. Assoc. Avan. Sci.* **54**, 250–2.

GOETHE, F. (1937). Beobachtungen und Untersuchen zur Biologie der Silbermöwe auf der Vogelinsel Memmersand. *J. Orn.* **85**, 1–119.

GOLDSMITH, T. H. & GRIFFIN, D. R. (1956). Further observations of homing Terns. *Biol. Bull.* **111**, 235–9.

GORDON, D. A. (1948). Sensitivity of the homing pigeon to the magnetic field of the earth. *Science*, **108**, 710–11.

GOULD, E. (1957). Orientation in Box Turtles, *Terrapene c. carolina* (Linnaeus). *Biol. Bull.* **112**, 336–48.

GOULD, E. (1960). Discussion. *Cold Spring Harbour Symp.* **25**, 461.

GRABER, R. R. & COCHRAN, W. W. (1959). An audio technique for the study of nocturnal migration of birds. *Wilson Bull.* **71**, 220–6.

GRABER, R. R. & COCHRAN, W. W. (1960). Evaluation of an aural record of nocturnal migration. *Wilson Bull.* **72**, 251–73.

GRABER, R. R. & HASSLER, S. S. (1962). The effectiveness of aircraft-type (APS) radar in detecting birds. *Wilson Bull.* **74**, 367–80.

GRANIT, R. (1959). Neural activity in the retina. In *Handbook of Physiology* (ed. J. Field), vol. 1, 693–712. Washington D.C.

GRAUE, L. C. (1963). The effect of phase shifts in the day-night cycle on pigeon homing at distances of less than one mile. *Ohio J. Sci.* **63**, 214–17.

GRAUE, L. C. (1965a). Experience effect on initial orientation in pigeon homing. *Anim. Behav.* **13**, 149–53.

GRAUE, L. C. (1965b). Initial orientation in pigeon homing related to magnetic contours. *Am. Zool.* **5**, 704.

GRAUE, L. C. & PRATT, J. G. (1959). Directional differences in pigeon homing in Sacramento, California and Cedar Rapids, Iowa. *Anim. Behav.* **7**, 201–8.

GREGORY, R. L. (1966). *Eye and Brain.* London.

GRIFFIN, D. R. (1940). Homing experiments with Leach's Petrels. *Auk*, **57**, 61–74.

GRIFFIN, D. R. (1943). Homing experiments with Herring Gulls and Common Terns. *Bird Band.* **14**, 7–33.

GRIFFIN, D. R. (1944). The sensory basis of bird navigation. *Quart. Rev. Biol.* **19**, 21–32.

GRIFFIN, D. R. (1952a). Radioactive tagging of animals under natural conditions. *Ecology*, **33**, 329–35.

GRIFFIN, D. R. (1952b). Airplane observations of homing pigeons. *Bull. Mus. Comp. Anat.* **107**, 411–40.

GRIFFIN, D. R. (1952c). Bird navigation. *Biol. Rev.* **27**, 359–400.

GRIFFIN, D. R. (1953). Acoustic orientation in the Oil Bird, *Steatornis*. *Proc. Nat. Acad. Sci.* **39**, 884–93.

GRIFFIN, D. R. (1955). Bird navigation. In *Recent Studies in Avian Biology*, (Ed. A. Wolfson). Urbana. 154–97.

GRIFFIN, D. R. (1958). *Listening in the Dark.* New Haven.

GRIFFIN, D. R. (1964). *Bird Migration.* New York.

GRIFFIN, D. R. & GOLDSMITH, T. H. (1955). Initial flight directions of homing birds. *Biol. Bull.* **108**, 264–76.

GRIFFIN, D. R. & HOCK, R. J. (1949). Airplane observations of homing birds. *Ecology*, **30**, 176–98.

GRINNELL, A. D. (1963). The neurophysiology of audition in bats: temporal parameters. *J. Physiol.* **167**, 67–96.

GRUNDLACH, R. H. (1932). A field study of homing pigeons. *J. Comp. Psychol.* **13**, 397–402.

GRUNDLACH, R. H. (1933). Visual acuity of homing pigeons. *J. Comp. Psychol.* **16**, 327–42.

GRUYS-CASIMIR, E. M. (1965). On the influence of environmental factors on the autumn migration of Chaffinch and Starling: a field study. *Arch. néerl. Zool.* **16**, 175–279.

GWINNER, E. (1966). Tagesperiodische schwankungen der Vorzughelligkeit bei Vogeln. *Z. vergl. Physiol.* **52**, 370–9.

HAARTMAN, L. v. (1960). The Orstreue of the Pied Flycatcher. *Proc. XII Int. Orn. Cong. Helsinki*, **1**, 266–73.

HACHET-SOUPLET, P. (1901). De la faculté de directioná grandes distances chez le pigeon voyageur et chez les animaux en général. *Ann. Psychol. Zool.* **1**, 22–6.

HACHET-SOUPLET, P. (1911). L'instinct du retour chez le pigeon voyageur. *Rev. sci. Paris*, **29**, 231–8.

HAMILTON, W. F. & GOLDSTEIN, J. L. (1933). Visual acuity and accommodation in the pigeon. *J. Comp. Psychol.* **15**, 193–7.

HAMILTON, W. J. (1962a). Celestial orientation in juvenal waterfowl. *Condor*, **64**, 19–33.

HAMILTON, W. J. (1962b). Initial orientation and homing of inexperienced Pintails. *Bird Band.* **33**, 61–9.

HAMILTON, W. J. (1962c). Does the Bobolink navigate? *Wilson Bull.* **74**, 357–66.

HAMILTON, W. J. (1962d). Bobolink migratory pathways and their experimental analysis under night skies. *Auk*, **79**, 208–33.

HAMILTON, W. J. (1966). Analysis of bird navigation experiments. In *Systems Analysis in Ecology* (ed. K. E. F. Watt), New York. 147–78.

HAMILTON, W. J. & HAMMOND, M. C. (1960). Oriented overland spring migration of pinioned Canada Geese. *Wilson Bull.* **72**, 385–91.

HARDY, E. (1951). *Pigeon Guide: A Complete Handbook of Pigeon Keeping.* London.

HARKER, J. E. (1964). *The Physiology of Diurnal Rhythms.* Cambridge.

HARPER, W. G. (1958). Detection of bird migration by centimetric radar—a cause of radar 'angels'. *Proc. Roy. Soc.* B, **149**, 484–502.

HASLER, A. D. & SCHWASSMANN, H. O. (1960). Sun orientation of fish at different latitudes. *Cold Spring Harbour Symp.* **25**, 429–41.

HASSLER, S. S., GRABER, R. R. & BELLROSE, F. C. (1963). Fall migration and weather, a radar study. *Wilson Bull.* **75**, 56–77.

HECHT, S. & PIRENNE, M. H. (1940). The sensibility of the nocturnal Long-eared Owl in the spectrum. *J. Gen. Physiol.* **23**, 709–17.

HEINROTH, O. & HEINROTH, K. (1941). Das Heimfinde-Vermögen der Brieftauben. *J. Orn.* **89**, 213–56.

HENDERSON, G. H. (1948). Physical basis of bird navigation. *Science*, **107**, 597–8.

HEYLAND, J. D. (1965). The orientation and homing of upland nesting waterfowl in southwestern Saskatchewan. *Ph.D. Thesis, Univ. Guelph.*

HILPRECHT, A. (1935). Heimfindeversuche mit Wintervögeln. *Vogelzug*, **6**, 188–96.

HITCHCOCK, H. B. (1952). Airplane observations of homing pigeons. *Proc. Amer. Phil. Soc.* **96**, 270–89.

HITCHCOCK, H. B. (1955). Homing flights and orientation in pigeons. *Auk*, **72**, 355–73.

HOCHBAUM, H. A. (1955). *Travels and Traditions of Waterfowl.* Minneapolis.

HOFFMANN, K. (1954). Versuche zu der im Richtungsfinden der Vögel enthaltenen Zeitschätzung. *Z. Tierpsychol.* **11**, 453–75.

HOFFMANN, K. (1958). Repetition of an experiment on bird orientation. *Nature, Lond.* **181**, 1435–7.

HOFFMANN, K. (1959a). Über den Einfluss verschiendener Faktoren auf die Heimkehrleistung von Brieftauben. *J. Orn.* **100**, 90–102.

HOFFMANN, K. (1959b). Die Richtungsorientierung von Staren unter der Mitternachtssonne. *Z. vergl. Physiol.* **41**, 471–80.

HOFFMANN, K. (1960a). Versuche zur Analyse der Tagesperiodik. I. Der Einfluss der Lichtintensität. *Z. vergl. Physiol.* **43**, 544–66.

HOFFMANN, K. (1960b). Experimental manipulation of the orientational clock in birds. *Cold Spring Harbour Symp.* **25**, 379–87.

HOFFMANN, K. (1965). Clock-mechanisms in celestial orientation of animals. In *Circadian Clocks* (ed. J. Aschoff), Amsterdam 426–41.

HORRIDGE, G. A. (1966a). Optokinetic memory in the Crab, *Carcinus*. *J. Exp. Biol.* **44**, 233–45.

HORRIDGE, G. A. (1966b). Optokinetic memory in the Locust. *J. Exp. Biol.* **44**, 255–61.

HORRIDGE, G. A. (1966c). Optokinetic responses of the Crab, *Carcinus* to a single moving light. *J. Exp. Biol.* **44**, 263–74.

HORRIDGE, G. A. (1966d). Direct response of the Crab, *Carcinus* to the movement of the sun. *J. Exp. Biol.* **44**, 275–83.

HOUGHTON, E. W. & LAIRD, A. G. (1967). A preliminary investigation into the use of radar as a deterrent of bird strikes on aircraft. *R.R.E. Memorandum* 2353.

HOWELL, J. C., LASKEY, A. R. & TANNER, J. T. (1954). Bird mortality at airport ceilometers. *Wilson Bull.* **66**, 207–15.

HUIZINGER, E. (1935). Durchschneidung aller Bogengänge bei der Taube. *Pflüg. Arch. ges. Physiol.* **236**, 52–8.

ISING, G. (1945). Die physikalische Möglichkeit eines tierischen Orientierungssinnes auf Basis der Erdrotation. *Ark. Mat. Astr. Fys.* **32**, 1–23.

JACOBS, A. A. M. C. (1962). Richtingkeuse van Spreeuw en Vink in de Kramer-kooi. *Limosa*, **35**, 187–92.

JOHNSTON, D. W. & HAINES, T. P. (1957). Analysis of mass bird mortality in October 1954. *Auk*, **74**, 447–58.

KALMUS, H. (1956). Sun navigation of *Apis mellifica* L. in the southern hemisphere. *J. Exp. Biol.* **33**, 554–65.

KENDALL, D. G. & SPEAKMAN, J. (1968). (personal communication).

KENYON, K. W. & RICE, D. W. (1958). Homing of Laysan Albatrosses. *Condor*, **60**, 3–6.

KIMM, I. H. (1960). Orientation of Cockchafers. *Nature, Lond.* **188**, 69–70.

KING, J. M. B. (1959). Orientation of migrants over sea in fog. *Brit. Birds*, **52**, 125–6.

KLUIJVER, H. W. (1935). Ergebnisse eines Versuches über das Heimfindevermögen von Staren. *Ardea*, **24**, 227–39.

KNIERIEM, H. (1943). Voraussetzungen für schnelles Heimfinden der Brieftauben bei geringen Verlusten auf den Reisen. *Z. Tierpsychol.* **5**, 131–52.

KNOOR, O. A. (1954). The effect of radar on birds. *Wilson Bull.* **66**, 264.

KOPYSTYŃSKA, K. (1962). Investigations on the vision of infrared in animals. III. Preliminary experiments on the Little Owl *Athene noctua* (Scop.). *Zesz. Nauk. U.J. Zoologia*, **7**, 95–107.

KRAMER, G. (1931). Zug in grosser Höhe. *Vogelzug*, **2**, 69–71.

KRAMER, G. (1948). Neue Beiträge zur Frage der Fernorientierung der Vögel. *Orn. Ber.* **1**, 228–39.

KRAMER, G. (1949). Über Richtungstendenzen bei der nächtlichen Zugunruhe gekäfigter Vögel. In *Ornithologie als biologische Wissenschaft.* Heidelberg.

KRAMER, G. (1951a). Eine neue Methode zur Erforschung der Zugorientierung und die bisher damit erzielten Ergebnisse. *Proc. X Int. Orn. Cong. Uppsala*, 271–80.

KRAMER, G. (1951b). Versuche zur Wahrnehmung von Ultrakurzwellen durch Vögel. *Vogelwarte*, **16**, 56–9.

KRAMER, G. (1952). Experiments on bird orientation. *Ibis*, **94**, 265–85.

KRAMER, G. (1953a). Wird die Sonnenhöhe bei der Heimfindeorientierung verwertet?. *J. Orn.* **94**, 201–19.

KRAMER, G. (1953b). Danebenfliegen un Überfliegen beim Heimflug von Brieftauben. *Vogelwarte*, **16**, 146–8.

KRAMER, G. (1953c). Die Sonnenorientierung der Vögel. *Verh. dtsch. Zool. Ges. Freiburg, 1952*, 72–84.

KRAMER, G. (1954). Einfluss von Temperatur und Erfahrung auf das Heimfindevermögen von Brieftauben. *J. Orn.* **95**, 343–7.

KRAMER, G. (1955). Ein weiterer Versuch, die Orientierung von Brieftauben durch jahreszeitliche Änderung der Sonnenhöhe zu beeinflussen. Gleichzeitig eine Kritik der Theorie des Versuchs. *J. Orn.* **96**, 173–85.

KRAMER, G. (1956). Über Flüge von Brieftauben über See. *Natur u. Jagd in Niedersachsen*, 113–17.

KRAMER, G. (1957). Experiments in bird orientation and their interpretation, *Ibis*. **99**, 196–227.

KRAMER, G. (1959). Recent experiments on bird orientation. *Ibis*, **101**, 399–416.

KRAMER, G. (1961). Long-distance orientation. In *Biology and Comparative Physiology of Birds* (ed. A. J. Marshall), London 341–71.

KRAMER, G., PRATT, J. G. & ST PAUL, U. v. (1956). Directional differences in pigeon homing. *Science*, **123**, 329–30.

KRAMER, G., PRATT, J. G. & ST PAUL, U. v. (1957). Two direction experiments with homing pigeons and their bearing on the problem of goal orientation. *Amer. Nat.* **91**, 37–48.

KRAMER, G., PRATT, J. G. & ST PAUL, U. v. (1958). Neue Untersuchungen über den 'Richtungseffekt'. *J. Orn.* **99**, 178–91.

KRAMER, G. & RIESE, E. (1952). Die Dressur von Brieftauben auf Kompassrichtung im Wahlkäfig. *Z. Tierspychol.* **9**, 245–51.

KRAMER, G. & ST PAUL, U. v. (1950). Ein Wesentlicher Bestandteil der Orientierung der Reisetaube: die Richtungsdressur. *Z. Tierpsychol.* **7**, 620–31.

KRAMER, G. & ST PAUL, U. v. (1952). Heimkehrleistungen von Brieftauben ohne Richtungsdressur. *Verh. dtsch. Zool. Ges. 1951*, 172–8.

KRAMER, G. ST PAUL, U. v. (1954). Das Heimkehvermögen gekafigter Brieftauben. *Orn. Beob.* **51**, 3–12.

KRAMER, G. & ST PAUL, U. v. (1956a). Weitere Erfahrungen über den 'Wintereffekt' beim Heimfindevermögen von Brieftauben. *J. Orn.* **97**, 353–70.

KRAMER, G. & ST PAUL, U. v. (1956b). Über das Heimfinden von Kafigtauben über Kurzstrecken. *J. Orn.* **97**, 371–6.

KRAMER, G. & ST PAUL, U. v. & WALLRAFF, H. G. (1958). Über die Heimfindeleistung von unter Sichtbegrenzung aufgewachsenen Brieftauben. *Verh. dtsch. Zool. Ges. Frankfurt, 1958*, 168–76.

KRAMER, G. & SEILKOPF, H. (1950). Heimkehrleistungen von Reisetauben in Abhangigkeit vom Wetter, insbesondere vom Wind. *Vogelwarte*, **15**, 242–7.

KRÄTZIG, H. & SCHÜZ, E. (1936). Ergebnis der Versetzung ostbaltischer Stare ins Binnenland. *Vogelzug*, **7**, 163–75.

Lack, D. (1943–44). The problem of partial migration. *Brit. Birds*, **37**, 122–30, 143–50.

Lack, D. (1958). Migrational drift of birds plotted by radar. *Nature, Lond.* **158**, 221–3.

Lack, D. (1959–63). Migration across the North Sea studied by radar: I. Survey throughout the year. II. The spring departure 1956–9. III. Movements in June and July. IV. Autumn. *Ibis*, **101**, 209–34; **102**, 26–57; **104**, 74–85; **105**, 1–54.

Lack, D. (1959). Migration across the sea. *Ibis*, **101**, 374–99.

Lack, D. (1960a). The height of bird migration. *Brit. Birds*, **53**, 5–10.

Lack, D. (1960b). Autumn 'drift migration' on the English east coast. *Brit. Birds*, **53**, 325–52, 379–97.

Lack, D. (1960c). A comparison of 'drift migration' at Fair Isle, the Isle of May and Spurn Point. *Scot. Birds*, **1**, 295–327.

Lack, D. (1960d). The influence of weather on passerine migration. A review. *Auk*, **77**, 171–209.

Lack, D. (1962). Radar evidence on migration orientation. *Brit. Birds*, **55**, 139–58.

Lack, D. & Eastwood, E. (1962). Radar films of migration over eastern England. *Brit. Birds*, **55**, 338–414.

Lack, D. & Lockley, R. M. (1938). Skokholm Bird Observatory homing experiments. I. 1936–37. Puffins, Storm Petrels and Manx Shearwaters. *Brit. Birds*, **31**, 242–8.

Lack, D. & Parslow, J. L. F. (1962). Falls of night migrants on the east coast in autumn 1960 and 1961. *Bird Migration*, **2**, 187–200.

Lack, D. & Varley, G. C. (1945). Detection of birds by radar. *Nature, Lond.* **156**, 446.

Landsberg, H. (1948). Bird migration and pressure patterns. *Science*, **108**, 708–9.

leFebvre, E. A., Birkebak, R. C. & Dorman, F. D. (1967). A flight-time integrator for birds. *Auk*, **84**, 124–8.

Lee, S. L. B. (1963). Migration in the Outer Hebrides studied by radar. *Ibis*, **105**, 493–515.

Leibowitz, H. W. (1955). The relation between rate threshold for the perception of movement and luminence for various durations of exposure. *J. exp. Psychol.* **49**, 209–14.

Levine, J. (1955). Consensual pupillary response in birds. *Science*, **122**, 690.

Libby, O. G. (1899). The nocturnal flight of migrating birds. *Auk*, **16**, 140–6.

Lincoln, F. C. (1927). The military use of the homing pigeon. *Wilson Bull.* **34**, 67–74.

Lindauer, M. (1957). Sonnenorientierung der Bienen unter der Äquatorsonne and zur Nachzeit. *Naturwiss.* **44**, 1–6.

Lindauer, M. (1959). Angeborene und erlernte Komponenten in der Sonnenorientierung der Bienen. Bermerkungen und Versuche zur einer Arbeit von Kalmus. *Z. vergl. Physiol.* **42**, 43–62.

Lissmann, H. W. (1958). On the function and evolution of electric organs in fish. *J. Exp. Biol.* **35**, 156–91.

Lissmann, H. W. & Machin, K. E. (1958). The mechanism of object location in *Gymnarchus niloticus* and similar fish. *J. Exp. Biol.* **35**, 451–86.

LOCKIE, J. D. (1952). Comparison of some aspects of the retinae of the Manx Shearwater, Fulmar Petrel and the House Sparrow. *Quart. Rev. Micr. Sci.* **93**, 347–56.

LOCKLEY, R. M. (1942). *Shearwaters.* London.

LÖHRL, H. (1959). Zur frage des Zeitpunktes einer Prägung auf die Heimatregion beim Halbandschnapper (*Ficedula albicollis*). *J. Orn.* **100**, 132–40.

LORD, R. D. (1956). A comparative study of the eyes of some falconiform and passeriform birds. *Am. Mid. Nat.* **56**, 325–44.

LORD, R. D., BELLROSE, F. C. & COCHRAN, W. W. (1962). Radiotelemetry of the respiration of a flying duck. *Science*, **137**, 39–40.

LOWENSTEIN, O. (1950). Labyrinth and equilibrium. *Symp. Soc. Exp. Biol.* **4**, 60–82.

LOWERY, G. H. (1951). A quantitative study of the nocturnal migration of birds. *Univ. Kan. Mus. Nat. Hist.* **3**, 361–472.

LOWERY, G. H. & NEWMAN, R. J. (1966). A continent-wide view of bird migration on four nights in October. *Auk*, **83**, 547–86.

MCCABE, R. A. (1947). The homing of transplanted young Wood Ducks. *Wilson Bull.* **59**, 104–9.

MCCABE, R. A. & HALE, J. B. (1960). An attempt to establish a colony of Yellow-headed Blackbirds. *Auk*, **77**, 425–32.

MCCANN, G. D. & MACGINITIE, G. F. (1965). Optomotor response studies of insect vision. *Proc. Roy. Soc.* B **163**, 369–401.

MCILHENNY, E. A. (1934). Twenty-two years of banding migratory wildfowl at Avery Island, Louisiana. *Auk*, **51**, 328–37.

MCILHENNY, E. A. (1940). An early experiment in the homing ability of wildfowl. *Bird Band.* **11**, 58.

MANWELL, R. D. (1941). Homing instinct of the Red-winged Blackbird. *Auk*, **58**, 185–7.

MANWELL, R. D. (1962). The homing of Cowbirds. *Auk*, **79**, 649–54.

MARSHALL, A. J. (1961). Breeding seasons and migrations. In *Biology and Comparative Physiology of Birds*, vol. 2, New York and London. 307–39.

MARSHALL, A. J. & SERVENTY, D. L. (1959). Experimental demonstration of an internal rhythm of reproduction in a trans-equatorial migrant (the Short-tailed Shearwater, *Puffinus tenuirostris*). *Nature, Lond.* **184**, 1704–5.

MARTORELLI, G. (1907). Di alcune nuove apparizioni in Italia di uccelli migratori siberiani ed americani e dell' influenza del moto rotatorie dell Terra sulla direzione generale delle migrazioni. *Atti Soc. ital. Sci. nat.* **46**, 1–30.

MASCHER, J. W., STOLT, B. O. & WALLIN, L. (1962). Migration in spring recorded by radar and field observations in Sweden. *Ibis*, **104**, 205–15.

MATTHEWS, G. V. T. (1948). Bird navigation. *New Nat.* **1**, 146–55.

MATTHEWS, G. V. T. (1951 a). The sensory basis of bird navigation. *J. Inst. Nav.* **4**, 260–75.

MATTHEWS, G. V. T. (1951 b). The experimental investigation of navigation in homing pigeons. *J. Exp. Biol.* **28**, 508–36.

MATTHEWS, G. V. T. (1952 a). An investigation of homing ability in two species of gulls. *Ibis*, **94**, 243–64.

MATTHEWS, G. V. T. (1952 b). The relation of learning and memory to the orientation and homing of pigeons. *Behaviour*, **4**, 202–41.

MATTHEWS, G. V. T. (1953a). Sun navigation in homing pigeons. *J. Exp. Biol.* **30**, 243–67.

MATTHEWS, G. V. T. (1953b). The orientation of untrained pigeons: a dichotomy in the homing process. *J. Exp. Biol.* **30**, 268–76.

MATTHEWS, G. V. T. (1953c). Navigation in the Manx Shearwater. *J. Exp. Biol.* **30**, 370–96.

MATTHEWS, G. V. T. (1954). Some aspects of incubation in the Manx Shearwater *Procellaria puffinus*, with particular reference to chilling resistance in the embryo. *Ibis*, **96**, 432–40.

MATTHEWS, G. V. T. (1955a). An investigation of the 'chronometer' factor in bird navigation. *J. Exp. Biol.* **32**, 39–58.

MATTHEWS, G. V. T. (1955b). *Bird Navigation*, 1st edition. Cambridge.

MATTHEWS, G. V. T. (1956). The sensory nature of bird navigation. *Ciba Found. Symp. on Extrasensory Perception*. London. 156–64.

MATTHEWS, G. V. T. (1959). Discussion on migration and orientation. *Ibis*, **101**, 427–8.

MATTHEWS, G. V. T. (1961). 'Nonsense' orientation in Mallard, *Anas platyrhynchos*, and its relation to experiments on bird navigation. *Ibis*, **103a**, 211–30.

MATTHEWS, G. V. T. (1962). Tests of the possible social significance of 'nonsense' orientation. *Wildfowl Trust Ann. Rep.* **13**, 47–52.

MATTHEWS, G. V. T. (1963a). 'Nonsense' orientation as a population variant. *Ibis*, **105**, 185–97.

MATTHEWS, G. V. T. (1963b). The orientation of pigeons as affected by the learning of landmarks and by the distance of displacement. *Anim. Behav.* **11**, 310–17.

MATTHEWS, G. V. T. (1963c). The astronomical bases of 'nonsense' orientation. *Proc. XIII Int. Orn. Cong. Ithaca*, 415–29.

MATTHEWS, G. V. T. (1964). Individual experience as a factor in the navigation of Manx Shearwaters. *Auk*, **81**, 132–46.

MATTHEWS, G. V. T. (1967). Some parameters of 'nonsense' orientation in Mallard. *Wildfowl Trust Ann. Rep.* **18**, 88–97.

MATTHEWS, G. V. T., EYGENRAAM, J. A. & HOFFMANN, L. (1963). Initial direction tendencies in the European Greenwinged Teal. *Wildfowl Trust Ann. Rep.* **14**, 120–3.

MATTHEWS, L. H. & MATTHEWS, B. H. C. (1939). Owls and infrared radiation. *Nature, Lond.* **143**, 983.

MATTINGLEY, A. H. E. (1946). Orientation in birds. *Ibis*, **88**, 512–17.

MATURANA, H. R. & FRENK, S. (1963). Directional movement and horizontal edge detectors in the pigeon retina. *Science*, **142**, 977–9.

MATURANA, H. R. & FRENK, S. (1965). Synoptic connections of the centrifugal fibres in the pigeon retina. *Science*, **150**, 359.

MAUERSBERGER, G. (1957). Umsiedlungsversuche am Trauerschnäpper (*Musicacapa hypoleuca*), durchgefährt in der Sowjetunion—ein Sammelreferat. *J. Orn.* **99**, 445–7.

MAY, W. E. (1950). The Double Altitude problem. *J. Inst. Nav.* **3**, 416–21.

MAYHEW, W. W. (1963). Homing of Bank Swallows and Cliff Swallows. *Bird Band.* **34**, 179–90.

MAZZEO, R. (1953). Homing of the Manx Shearwater. *Auk*, **70**, 200–1.

MEDWAY, Lord (1959). Echo-location among *Collocalia*. *Nature, Lond.* **184**, 1352–3.

MEISE, W. (1933). Kinaesthetisches Gedächtnis und Fernorientierung der Vögel. *Vogelzug*, **4**, 101–13.

MELLO, N. K. (1965). Interhemispheric reversal of mirror-image oblique lines after monocular training in pigeons. *Science*, **148**, 252–4.

MENNER, E. (1938). Die Bedeutung des Pecten im Auge des Vogels für die Wahrnehmung von Bewegungen. *Zool. Jb. Allg. Zool.* **58**, 481–538.

MERKEL, F. W. & FROMME, H. G. (1958). Untersuchungen über das Orientierungsvermögen nächtlich ziehender Rotkehlchen (*Erithacus rubecula*). *Naturwiss.* **45**, 499–500.

MERKEL, F. W., FROMME, H. G. & WILTSCHKO, W. (1963). Nichtvisuelles Orientierungsvermögen bei nächtlich zugunruhigen Rotkehlchen! *Vogelwarte*, **22**, 168–73.

MERKEL, F. W. & WILTSCHKO, W. (1965). Magnetismus und Richtungsfinden zugunruhiger Rotkehlchen (*Erithacus rubecula*). *Vogelwarte*, **23**, 71–7.

MERKEL, F. W. & WILTSCHKO, W. (1966). Nächtliche Zugunruhe und Zugorientierung bei Kleinvögeln. *Verh. dtsch. Zool. Ges. Jena*, 356–61.

MEWALDT, L. R. (1963). Californian 'crowned' Sparrows return from Louisiana. *Western Bird Bander*, **38**, 1–4.

MEWALDT, L. R. (1964). California Sparrows return from displacement to Maryland. *Science*, **146**, 941–2.

MEWALDT, L. R. & FARNER, D. S. (1957). Translocated Golden-crowned Sparrows return to winter range. *Condor*, **59**, 268–9.

MEWALDT, L. R., MORTON, M. L. & BROWN, I. L. (1964). Orientation of migratory restlessness in *Zonotrichia*. *Condor*, **66**, 377–417.

MEWALDT, L. R. & ROSE, R. G. (1960). Orientation of migratory restlessness in the White-crowned Sparrow. *Science*, **131**, 105–6.

MEYER, A. (1938). *Z. Brieftaubenkunde*, **38**, 7–8. (*Vogelzug*, **9**, 39.)

MEYER, M. E. (1964). Discriminative basis for astronavigation in birds. *J. Comp. Physiol. Psychol.* **58**, 403–6.

MEYER, M. E. (1966a). The internal clock hypothesis for astronavigation in homing pigeons. *Psychon. Sci.* **5**, 259–60.

MEYER, M. E. (1966b). Sensitivity of the pigeon to changes in the magnetic field. *Psychon. Sci.* **5**, 349–60.

MICHENER, M. C. & WALCOTT, C. (1967). Homing of single pigeons—an analysis of tracks. *J. Exp. Biol.* **47**, 99–131.

MIDDENDORF, A. v. (1855). Die Isepipetsen Russlands; Grundlagen zur Erforschung der Zugzeiten und Zugrichtungen der Vögel Russlands. *Mem. Acad. Sci. St. Petersbourg*, **8**, 1–143.

MITTELSTAEDT, H. (1962). Control systems of orientation in insects. *Ann. Rev. Entomology*, **7**, 177–98.

MITTELSTAEDT, H. (1964). Basic control patterns of orientational homeostasis. *Symp. Soc. Exp. Biol.* **18**, 365–85.

MONTGOMERY, K. C. & HEINEMANN, E. G. (1952). Concerning the ability of homing pigeons to discriminate patterns of polarised light. *Science*, **116**, 454–6.

MOOK, J. H., ROTH, J. & ZILJLSTRA, J. J. (1957). Stichting Vogeltrekstation

Texel 1956. De Vogeltrekwaarnemmgen op de Noord-Veluwe. *Limosa*, **30**, 62–75.

MOORE, J. (1735). *Columbarium: or the Pigeon House*. London.

MOREAU, R. E. (1961). Problems of Mediterranean–Saharan migration. *Ibis*, **103**a, 373–427, 580–623.

MURPHY, J. J. (1873). Instinct. A mechanical analogy. *Nature, Lond.* **7**, 483.

MYRES, M. T. (1964). Dawn ascent and re-orientation of Scandinavian thrushes (*Turdus* spp.) migrating at night over the northeastern Atlantic Ocean. *Ibis*, **106**, 7–51.

NEW, D. A. T. & NEW, J. K. (1962). The dances of honey bees at small zenith distances of the sun. *J. Exp. Biol.* **39**, 271–91.

NEWMAN, J. & LOWERY, G. H. (1962). The study of the ultrahigh daytime migration with a spotting 'scope'. *Proc. XIII Int. Orn. Cong. Ithaca.* (unpublished paper).

NICE, M. M. (1933). Migratory behaviour in Song Sparrows. *Condor*, **35**, 219–24.

NICOL, J. A. C. (1945). The homing ability of the Carrier Pigeon: its value in warfare. *Auk*, **62**, 286–98.

NISBET, I. C. T. (1959). Calculation of flight directions of birds observed crossing the face of the moon. *Wilson Bull.* **71**, 237–43.

NISBET, I. C. T. (1963a). Measurements with radar of the height of nocturnal migration over Cape Cod, Massachusetts. *Bird Band.* **34**, 57–67.

NISBET, I. C. T. (1963b). Quantitative study of migration with 23-centimetre radar. *Ibis*, **105**, 435–60.

NISBET, I. C. T. & DRURY, W. H. (1967). Orientation of spring migrants studied by radar. *Bird Band.* **38**, 173–86.

NOBLE, C. E. (1949). The perception of the vertical. III. The visual vertical as a function of centrifugal and gravitational forces. *J. Exp. Psychol.* **39**, 839–50.

ODUM, H. T. (1948). The bird navigation controversy. *Auk*, **65**, 584–97.

OEHME, H. (1962). Das Auge von Mauersegler, Star und Amsel. *J. Orn.* **103**, 187–212.

OORDT, G. J. V. & BOLS, C. J. A. C. (1929). Zum Orientierungsproblem der Vögel. Kastrationversuche an Brieftauben. *Biol. Zbl.* **49**, 173–86.

ORGEL, A. R. & SMITH, J. C. (1954). Test of the magnetic theory of homing. *Science*, **120**, 891–2.

OSMAN, W. H. (1950). *Pigeons in World War II*. London.

PALMEN, J. A. (1876). *Über die Zugstrassen der Vögel*. Leipzig.

PAPI, F. & PARDI, L. (1953). Ricerche sull'orientamento di *Talitrus saltator* (Montagu) (Crustacea-Amphipoda). *Z. vergl. Physiol.* **35**, 459–518.

PAPI, F. & PARDI, L. (1959). Nuovi reperti sull'orientamento lunare di *Talitrus saltator* Montagu (Crustacea-Amphipoda). *Z. vergl. Physiol.* **41**, 583–96.

PAPI, F. & PARDI, L. (1963). On the lunar orientation of sandhoppers (Amphipoda Talitridae). *Biol. Bull.* **124**, 97–105.

PAPI, F., SERRETTI, L. & PARRINI, S. (1957). Nuove richerche sull'orientamento e il senso del tempo di *Arctosa perita* (Latr.) (Araneae Lycosidae). *Z. vergl. Physiol.* **39**, 531–61.

PAPI, F. & SYRJÄMÄKI, J. (1963). The sun-orientation rhythm of Wolf Spiders at different latitudes. *Arch. ital. Biol.* **101**, 59–77.

PAPI, F. & TONGIORGI, P. (1963). Innate and learned components in the astronomical orientation of Wolf Spiders. *Ergebn. Biol.* **26**, 259–80.

PARDI, L. (1954). Esperienze sull'orientamento di *Talitrus saltator* (Montagu) (Crustacea-Amphipoda): l'orientamento al sole degli individui a ritmo nictiemerale invertito durante la 'loro notto'. *Boll. Ist. Mus. Zool. Univ. Torino*, **4**, 127.

PARDI, L. (1958). Esperienze sull'orientamento solare di *Phaleria provincialis* Fauv. (Coleopt.): Il compartamento a luce artificiale durante l'intero ciclo di 24 ore. *Atti Acc. Sc. Torino Cl. Sc. Fis. Mat.* **92**, 65.

PARSLOW, J. L. F. (1962). Immigration of night-migrants into southern England in spring 1962. *Bird Migration*, **2**, 160–76.

PATON, D. N. (1928). Reflex postural adjustments of balance in the duck. *Proc. Roy. Soc. Edinburgh*, **48**, 28–36.

PATTEN, B. C. (1964). The rational decision process in salmon migration. *J. Conseil Internat. p. l'Exploration de la Mer*, **28**, 410–17, 443–4.

PENNEY, R. L. & EMLEM, J. T. (1967). Further experiments on distance navigation in the Adelie Penguin. *Ibis*, **109**, 99–109.

PENNYCUICK, C. J. (1960a). The physical basis of astronavigation in birds: theoretical considerations. *J. Exp. Biol.* **37**, 573–93.

PENNYCUICK, C. J. (1960b). Sun navigation by birds. *Nature, Lond.* **188**, 1128.

PENNYCUICK, C. J. (1961). Sun navigation in birds? *Nature, Lond.* **190**, 1026.

PERDECK, A. C. (1957). Stichting vogeltrekstation Texel Jaarverslag over 1956. *Limosa*, **30**, 62–75.

PERDECK, A. C. (1958). Two types of orientation in migrating starlings *Sturnus vulgaris* L. and chaffinches *Fringilla coelebs* L., as revealed by displacement experiments. *Ardea*, **46**, 1–37.

PERDECK, A. C. (1960). Verplaatsingsproeven met Wintertaling en Spreeuw. *Limosa*, **33**, 75–8.

PERDECK, A. C. (1961). De standaardrichting van de Standinaafse vink. *Limosa*, **34**, 177–94.

PERDECK, A. C. (1962). De waarnemingen over de Trekrichting van de Vink in het buitenland. *Limosa*, **35**, 193–8.

PERDECK, A. C. (1963). Does navigation without visual clues exist in Robins? *Ardea*, **51**, 91–104.

PERDECK, A. C. (1964). An experiment on the ending of autumn migration in starlings. *Ardea*, **52**, 133–9.

PERDECK, A. C. (1967a). Oriëntatievermogen van Kokmeeuven. *Limosa*, **40**, 158–60.

PERDECK, A. C. (1967b). Orientation of starlings after displacement to Spain. *Ardea*, **55**, 194–202.

PETERSEN, E. (1953). Orienteringsforsøg med Haettemåge (*Larus r. ridibundus* (L.) og Stormmåge (*Larus c. canus* L.) i vinterkvarteret. *Dansk. Orn. Foren. Tidsskr.* **47**, 133–78.

PHILLIPS, J. H. (1963). The pelagic distribution of the Sooty Shearwater, *Procellaria grisea. Ibis*, **105**, 340–53.

PINOWSKI, J. (1967). Experimental studies on the dispersal of young Tree Sparrows. *Ardea*, **55**, 241–8.

PIRENNE, M. H. (1948). *Vision and the Eye*. London.

PITTENDRIGH, C. S. (1960). Circadian rhythms and the circadian organisation of living systems. *Cold Spring Harbour Symp.* **25**, 159–84.

PLATT, C. S. & DARE, R. S. (1945). The homing instinct in pigeons. *Science*, **101**, 439–40.

POLYAK, S. (1957). *The Vertebrate Visual System*. Chicago.

POOR, H. H. (1946). Birds and radar. *Auk*, **63**, 631.

PRATT, J. G. (1953). The homing problem in pigeons. *J. Parapsychol.* **17**, 34–60.

PRATT, J. G. (1955). An investigation of homing ability in pigeons without previous homing experience. *J. Exp. Biol.* **32**, 70–83.

PRATT, J. G. (1956). Testing for an ESP factor in pigeon homing. *Ciba Found. Symp. on Extrasensory Perception*, London 165–79.

PRATT, J. G. & THOULESS, R. H. (1955). Homing orientation in pigeons in relation to opportunity to observe the sun before release. *J. Exp. Biol.* **32**, 140–57.

PRATT, J. G. & WALLRAFF, H. G. (1958). Zwer- Richtungs-Versuche mit Brieftauben: Langstreckenflüge auf der Nord-Süd-Achse in Westdeutschland. *Z. Tierpsychol.* **15**, 332–9.

PRECHT, H. (1956). Einige Versuche zum Heimfindevermögen von Vögeln. *J. Orn.* **97**, 377–83.

PUMPHREY, R. J. (1948). The theory of the fovea. *J. Exp. Biol.* **25**, 299–312.

PUMPHREY, R. J. (1960). Sun navigation by birds. *Nature, Lond.* **188**, 1127.

PUMPHREY, R. J. (1961). Sensory organs: vision. In *Biology and Comparative Physiology of Birds* (ed. A. J. Marshall) II, pp. 55–68. London & New York.

PÜTZIG, P. (1938). Über das Zugverhalten umgesiedelter englischer Stockenten. *Vogelzug*, **9**, 139–45.

PYE, J. D. (1963). Mechanisms of echolocation. *Ergebn. Biol.* **26**, 12–20.

RANDIC, L. (1956). A device to determine position rapidly without calculation. *J. Inst. Nav.* **9**, 11–16.

RAWSON, K. S. & RAWSON, A. M. (1955). The orientation of homing pigeons in relation to change in sun declination. *J. Orn.* **96**, 168–72.

RENNER, M. (1959). Über ein weiteres versetzungsexperiment zur analyse des zeitsinnes und der Sonnenorientierung der honigbiene. *Z. vergl. Physiol.* **42**, 449–83.

RENNIE, J. (1835). *The Faculties of Birds*. London.

RHINE, J. B. (1951). The present outlook on the question of *psi* in animals. *J. Parapsychol.* **15**, 230–51.

RICHDALE, L. A. (1963). Biology of the Sooty Shearwater, *Puffinus griseus*. *Proc. Zool. Soc. Lond.* **141**, 1–117.

RIPER, W. v. & KALMBACH, E. R. (1952). Homing not hindered by wing magnets. *Science*, **115**, 577–8.

RIVIÈRE, B. B. (1923). Homing pigeons and pigeon racing. *Brit. Birds*, **17**, 118–38.

RIVIÈRE, B. B. (1929). The 'homing faculty' in pigeons. *Verh. VI Orn. Kongr. Copenhagen*, 535–55.

ROADCAP, R. (1962). Translocations of White-crowned and Golden-crowned Sparrows. *Western Bird Bander*, **37**, 55–7.

ROBERTS, T. W. (1942). Behaviour of organisms. *Ecol. Monogr.* **12**, 339–412.

ROBINSON, D. N. (1967). Visual discrimination of temporal order. *Science*, **156**, 1263–4.

ROCHON-DUVIGNEAUD, A. & MAURAIN, C. (1923). Enquête sur l'orientation du pigeon voyageur et son mécanisme. *La Nature*, **51**, 232–8.

ROWAN, M. K. (1952). The Greater Shearwater *Puffinus gravis* at its breeding grounds. *Ibis*, **94**, 97–121.

ROWAN, W. (1946). Experiments in bird migration. *Trans. Roy. Soc. Canada*, **40**, 123–35.

RÜPPELL, W. (1934*a*). Verfrachtungsversuche am Star (*Sturnus vulgaris*) u. a. Arten von W. Schein-Winsen. *Vogelzug*, **5**, 53–9.

RÜPPELL, W. (1934*b*). Heimfinde-Versuche mit Rauchschwalben (*Hirundo rustica*) und Mehlschwalben (*Delichon urbica*) von H. Warnat (Berlin-Charlottenburg). *Vogelzug*, **5**, 161–6.

RÜPPELL, W. (1935). Heimfindeversuche mit Staren 1934. *J. Orn.* **83**, 462–524.

RÜPPELL, W. (1936). Heimfindeversuche mit Staren und Schwalben 1935. *J. Orn.* **84**, 180–98.

RÜPPELL, W. (1937). Heimfindeversuche mit Staren, Rauchschwalben, Wendhälsen, Rotruckwürgen und Habichten 1936. *J. Orn.* **85**, 120–35.

RÜPPELL, W. (1938). Ergebnis eines Heimfindeversuches mit aufgezogen Staren. *Vogelzug*, **9**, 18–22.

RÜPPELL, W. (1940). Neue Ergebnisse über Heimfinden beim Habicht. *Vogelzug*, **11**, 57–64.

RÜPPELL, W. (1944). Versuche über Heimfinden ziehender Nebelkrähen nach Verfrachtung. *J. Orn.* **92**, 106–33.

RÜPPELL, W. & SCHEIN, W. (1941). Über das Heimfinden freilebender Stare bei Verfrachtung nach einjähriger Freihetsentziehung am Heimatort. *Vogelzug*, **12**, 49–56.

RÜPPELL, W. & SCHIFFERLI, A. (1939). Versuche über Winter Ortstreue an *Larus ridibundus* und *Fulica atra*, 1935. *J. Orn.* **87**, 224–39.

RÜPPELL, W. & SCHÜZ, E. (1948). Ergebnis der Verfrachtung von Nebelkrähen (*Corvus corone cornix*) wahrend des Wegzuges. *Vogelwarte*, **1**, 30–6.

ST PAUL, U. v. (1953). Nachweis der Sonnenorientierung bei nächtlich ziehenden Vögeln. *Behaviour*, **6**, 1–7.

ST PAUL, U. v. (1956). Compass directional training of Western Meadowlarks (*Sturnella neglecta*). *Auk*, **73**, 203–10.

ST PAUL, U. v. (1962). Das Nachtfliegen von Brieftauben. *J. Orn.* **103**, 337–43.

SAILA, S. B. & SHAPPY, R. A. (1963). Random movement and orientation in salmon migration. *J. Conseil. Internat. p. l'Exploration de la Mer*, **28**, 154–66, 440–3.

SARGENT, T. D. (1959). Winter studies of the Tree Sparrow, *Spizella arborea*. *Bird Band.* **30**, 27–37.

SARGENT, T. D. (1962). A study of homing in the Bank Swallow (*Riparia riparia*). *Auk*, **79**, 234–46.

SAUER, E. G. F. (1957). Die Sternenorientierung nächtlich ziehender Grasmücken. (*Sylvia atricapilla, borin* und *curruca*). *Z. Tierpsychol.* **14**, 29–70.

SAUER, E. G. F. (1961). Further studies on the stellar orientation of nocturnally migrating birds. *Psychol. Forschung.* **26**, 224–44.

SAUER, E. G. F. (1963). Migration habits of Golden Plovers. *Proc. XIII Int. Orn. Cong. Ithaca*, 454–67.

SAUER, E. G. F. & SAUER, E. M. (1955). Zur Frage der nächtlichen Zugorientierung von Grasmücken. *Rev. Suisse Zool.* **62**, 250–9.

SAUER, E. G. F. & SAUER, E. M. (1959). Nächtliche Zugorientierung europäischer Vögel in Südwestafrika. *Vogelwarte*, **20**, 4–31.

SAUER, E. G. F. & SAUER, E. M. (1960). Star navigation of nocturnal migrating birds. The 1958 planetarium experiments. *Cold Spring Harbour Symp.* **25**, 463–73.

SAUER, E. G. F. & SAUER, E. M. (1962). Richtungsfinden und Raumbeherrschung von Zugvögeln nach Gestirnen. *Nach. d. Obers-gessellschaft Bremen*, **50**, 12–16.

SCHAEFER, G. W. (1968). Bird recognition by radar. *Inst. Biol. Symp.* (in the Press).

SCHIFFERLI, A. (1936). Transportversuche mit Futterplatzvögeln im Herbst und Winter. *Orn. Beob.* **34**, 1.

SCHIFFERLI, A. (1942). Verfrachtungversuche mit Alpenseglern (*Micropus m. melba*) Solothurn-Lissabon. *Orn. Beob.* **39**, 145–50.

SCHIFFERLI, A. (1943a). Nachtrag zu Transportversuchen mit Futterplatzvögeln im Herbst und Winter. *Orn. Beob.* **40**, 43.

SCHIFFERLI, A. (1943b). Brandente (*Tadorna tadorna*) kehrt nach Sempach zurück. *Orn. Beob.* **40**, 44.

SCHIFFERLI, A. (1951). Transportversuche mit Alpenseglern (*Apus melba*) nach Nairobi. *Orn. Beob.* **48**, 183–4.

SCHMIDT-KOENIG, K. (1958). Experimentelle Einflussnahme auf die 24-Stunden-Periodik bei Brieftauben und deren Auswirkungen unter besonderer Berücksichtigung des Heimfindevermögens. *Z. Tierpsychol.* **15**, 301–31.

SCHMIDT-KOENIG, K. (1961a). Sun navigation in birds? *Nature, Lond.* **190**, 1025–6.

SCHMIDT-KOENIG, K. (1961b). Die Sonne als Kompass im Heim-Orientierungssystem der Brieftauben. *Z. Tierpsychol.* **18**, 221–44.

SCHMIDT-KOENIG, K. (1961c). Die Sonnenorientierung richtungdressierter Tauben in ihrer physiologischen Nacht. *Naturwiss.* **48**, 110.

SCHMIDT-KOENIG, K. (1963a). Hormones and homing in pigeons. *Physiol. Zool.* **36**, 264–72.

SCHMIDT-KOENIG, K. (1963b). Sun compass orientation of pigeons upon equatorial and trans-equatorial displacement. *Biol. Bull.* **124**, 311–21.

SCHMIDT-KOENIG, K. (1963c). On the role of the loft, the distance and site of release in pigeon homing (the 'cross loft' experiment). *Biol. Bull.* **125**, 154–64.

SCHMIDT-KOENIG, K. (1963d). Sun compass orientation of pigeons upon displacement north of the arctic circle. *Biol. Bull.* **127**, 154–8.

SCHMIDT-KOENIG, K. (1963e). Neuere Aspekte über die Orientierungsleistungen von Brieftauben. *Ergebn. Biol.* **26**, 286–97.

SCHMIDT-KOENIG, K. (1965a). Current problems in bird orientation. In *Advances in the Study of Behaviour* (ed. E. Hinde, K. Lehrman & E. Shaw), New York 217–78.

SCHMIDT-KOENIG, K. (1965b). Über den zeitlichen Ablauf der Angangsorientierung bei Brieftauben (Kurzfassurs). *Verh. dtsch. Zoll. Ges. 1964*, 409–11.

SCHMIDT-KOENIG, K. (1966). Über die Entfernung als Parameter bei der Anfangsorientierung der Brieftaube. *Z. vergl. Physiol.* **52**, 33–5.

SCHNEIDER, F. V. (1963). Ultraoptische Orientierung des Maikäfers (*Melolntha vulgaris* F.) in künstlichen elektrischen und magnetischen Feldern. *Ergebn. Biol.* **26**, 147–57.

SCHNEIDER, G. H. (1906). Die Orientierung der Brieftauben. *Z. Psychol. Physiol. Sinnesorg.* **40**, 252–79.

SCHREIBER, B., GUALTIEROTTI, T. & MAINARDI, D. (1962). Some problems of cerebellar physiology in migratory and sedentary birds. *Anim. Behav.* **10**, 42–7.

SCHULER, M. (1923). Die Störung von Pendul- und Kreiselapparaten durch bie Beschleunigung des Fahrzeuges. *Phys. Z.* **24**, 344–50.

SCHUMACHER, W. C. (1949). A preliminary study of a physical basis of bird navigation. *J. Appl. Phys.* **20**, 123.

SCHÜZ, E. (1938a). Über künstliche Verpflanzung bei Vögeln. *Proc. IX Cong. Orn. Int. Rouen*, 315–18.

SCHÜZ, E. (1938b). Auflassung ostpreussischer Jungstörche in England 1936. *Vogelzug*, **9**, 65–70.

SCHÜZ, E. (1949). Die Spät-Auflassung ostpreussischer Jungstörche in West-Deutschland durch die Vogelwarte Rossitten 1933. *Vogelwarte*, **15**, 63–78.

SCHÜZ, E. (1950). Früh-Auflassung ostpreussischer Jungstörche in West-Deutschland durch die Vogelwarte Rossitten 1933–36. *Bonner zool. Beitr.* **1**, 239–53.

SCHÜZ, E. (1952). *Vom Vogelzug: Grundriss der Vogelzugskunde*. Frankfurt.

SCHWARTZKOPFF, J. (1950). Sur Frage des 'Wahrnehmens' von Ultrakurzwellen durch Zugvögel. *Vogelwarte*, **15**, 194–6.

SCHWASSMANN, H. O. (1960). Environmental cues in the orientation rhythm of fish. *Cold Spring Harbour Symp.* **25**, 443–50.

SCHWASSMANN, H. O. & BRAEMER, W. (1961). The effect of experimentally changed photoperiod on the sun-orientation rhythm of fish. *Physiol. Zool.* **34**, 273–86.

SCHWASSMANN, H. O. & HASLER, A. D. (1963). The role of the sun's altitude in sun orientation of fish. *Physiol. Zool.* **37**, 163–78.

SCOTT, W. E. D. (1881). Some observations on the migrations of birds. *Bull. Nuttal Orn. Club*, **6**, 97–100.

SERVENTY, D. L. (1963). Egg-laying timetable of the Slender-billed Shearwater, *Puffinus tenuirostris*. *Proc. XIII Int. Orn. Cong. Ithaca*, 338–43.

SHARPE, A. (1963). *Ancient Voyagers in Polynesia*. London.

SHUMAKOV, M. E. (1965). Preliminary results of the investigation of migrational orientation of passerine birds by the round-cage method. *Bionica*, Moscow. 371–8 (in Russian).

SKINNER, B. F. (1950). Are theories of learning necessary? *Psychol. Rev.* **57**, 193–216.

SLADEN, W. J. L. & OSTENSO, N. A. (1960). Penguin tracks far inland in the Antarctic. *Auk*, **77**, 466–9.

SLEPIAN, J. (1948). Physical basis of bird navigation. *J. Appl. Phys.* **19**, 306.

SOBOL, E. D. (1930). Orienting ability of carrier pigeons with injured labyrinths. *Milit.-med. Ž. U.S.S.R.* **1**, 75. (*Biol. Abstr.* **8**, 15425.)

SOLLBERGER, A. (1965). *Biological Rhythm Research.* Amsterdam.

SOUTHERN, W. E. (1959). Homing of Purple Martins. *Wilson Bull.* **71**, 254–61.

SOUTHERN, W. E. (1964). Additional observations on winter Bald Eagle populations: including remarks on biotelemetry techniques and immature plumages. *Wilson Bull.* **76**, 121–37.

SOWLS, L. K. (1955). *Prairie Ducks.* Harrisburg, Pa.

SPAEPEN, J. & DACHY, P. (1952). Le problème de l'orientation chez les oiseaux migrateurs. II. Expériences préliminaires effectuées sur des Martinets noirs, *Apus apus* L. *Gerfaut*, **42**, 54–9.

SPAEPEN, J. & DACHY, P. (1953). Het Oriëntatieprobleem bij de Trekvogels. III. Verdere homingproeven met Gierzwaluwen (*Apus apus* L.). *Gerfaut*, **43**, 327–32.

SPALDING, D. A. (1873). Instinct. With original observations on young animals. *MacMillan's Magazine*, **27**, 282–93.

SPENCE, K. W. (1934). Visual acuity and its relation to brightness in chimpanzee and man. *J. Comp. Psychol.* **18**, 333–61.

STEIN, H. (1951). Untersuchungen über den Zeitsinn bei Vögeln. *Ž. vergl. Physiol.* **33**, 387–403.

STEWART, O. J. A. (1957). A bird's inborn navigational device. *Trans. Ky. Acad. Sci.* **18**, 78–84.

STIMMELMAYR, AL. (1930). Neue Wege zur Erforschung des Vogelzuges. *Verh. Orn. Gen. Bayern*, **19**, 149–85.

STOLPE, M. & ZIMMER, K. (1939). Der Schirrfluge der Kolibris im Zeitlupen-film. *J. Orn.* **87**, 136–55.

STONER, D. (1941). Homing instinct of the Bank Swallow. *Bird Band.* **12**, 104–9.

STRESEMANN, E. (1935). Haben die Vögel einen Ortssinn? *Ardea*, **24**, 213–26.

STRÖMBERG, G. (1961). Undersökning av nattsträcket i östra Blekinge. *Vår Fågelvärld*, **20**, 30–42.

SUFFERN, C. (1949). Pressure patterns in bird migration. *Science*, **109**, 209.

SUTTER, E. (1957). Radar als Hilfsmittel der Vogelzugforschung. *Orn. Beob.* **54**, 70–96.

SVÄRDSON, G. (1953). Visible migration within Fenno-Scandia. *Ibis*, **95**, 181–211.

TALKINGTON, L. (1967). Bird navigation and geomagnetism (Abstract). *Amer. Zoologist*, **7**, 199.

TANNER, J. A. (1966). Effect of microwave radiations on birds. *Nature, Lond.* **210**, 636.

TANSLEY, K. (1965). *Vision in Vertebrates.* London.

TEDD, J. G. & LACK, D. (1958). The detection of bird migration by high-power radar. *Proc. Roy. Soc.* B **149**, 503–10.

TEGETMEIER, W. H. (1871). *The Homing or Carrier Pigeon (Le Pigeon Voyageur). Its History, General Management and Method of Training.* London.

TETTENBORN, W. (1943). Festsellungen an beringten Lachmöwen in Berlin, Winter 1942/43. *J. Orn.* **91**, 286–95.

THAUZIÉS, A. (1910). L'orientation lointaine. *VI Int. Congr. Psychol.* 263–80, 834–5.

THAUZIÉS, A. (1913). L'orientation lointaine des pigeons voyageurs. *Rev. Sci.* **31**, 805–8.

THOMSON, A. L. (1926). *Problems of Bird-Migration.* London.

THOMSON, A. L. (1947). Scissors and paste are mightier than the pen. *Ibis,* **89**, 362–4.

THORPE, W. H. (1963*a*). *Learning and Instinct in Animals.* London.

THORPE, W. H. (1963*b*). Antiphonal singing in birds as evidence for avian auditory reaction. *Nature, Lond.* **197**, 774–6.

THORPE, W. H. & WILKINSON, D. H. (1946). Ising's theory of bird orientation. *Nature, Lond.* **158**, 903.

TINBERGEN, L. (1956). Field observations of migration and their significance for the problems of navigation. *Ardea,* **44**, 231–5.

TOMLINSON, R. E., WRIGHT, H. M. & HASKETT, T. S. (1960). Migrational homing, local movement and mortality of mourning doves in Missouri. *Trans. N. Amer. Wildl. Conf.* **25**, 253–67.

TONGIORGI, P. (1959). Effects of the reversal of the rhythm of nycthemeral illumination on astronomical orientation and diurnal activity in *Arctosa variana* C. L. Koch (Araneae-Lycosidae). *Arch. ital. Biol.* **97**, 251–65.

TREAT, A. E. (1947). The homing of pigeons following decompression to an indicated altitude of 25,000 feet. *Biol. Rev. Coll. City. N.Y.* **9**, 30–4. (*Biol. Abstr.* 1947, **21**, 16401.)

TRENDELENBURG, W. (1906). Über die Bewegung der Vögel nach Durchschneidung hinterer Rückenmarkswurzeln. *Arch. Anat. Physiol. Lpz.* (*Physiol.*), 1–126.

TUNMORE, B. G. (1960). A contribution to the theory of bird navigation. *Proc. XII Int. Orn. Cong. Helsinki,* 718–23.

VALIKANGAS, I. (1933). Finnische Zugvögel aus englischer Vögeleiern. *Vogelzug,* **4**, 159–66.

VANDERPLANK, F. L. (1934). The effect of infra-red waves on Tawny Owls (*Strix aluco*). *Proc. Zool. Soc. Lond.* 505–7.

VAUGHT, R. W. (1964). Results of transplanting flightless young Bluewinged Teal. *J. Wildlife Manag.* **28**, 208–12.

VAUGIEN, L. & VAUGIEN, M. (1963). La Moineau domestique synchronise le milieu de sa période alimentaire avec le milieu de la période journalière de lumière artificielle. *C.R. L'Acad. Sci.* **257**, 2040–2.

VIGUIER, C. (1882). Le sens d'orientation et ses organes chez les animaux et chez l'homme. *Rev. Phil.* **14**, 1–36.

VITALI, G. (1912). Di un interessante derivato dell' ectoderma della prima fessura branchiale nel passero. Un organo nervoso di senso nell' orecchio medio degli ucelli. *Anat. Anz.* **40**, 631–9.

VLEUGEL, D. A. (1953). Über die wahrscheinliche Sonnen-Orientierung einiger Vogelarten auf dem Zuge. *Orn. Fenn.* **30**, 41–51.

VLEUGEL, D. A. (1959). Über die wahrscheinlichste Method der Wind-Orientierung ziehender Buchfinken. *Orn. Fenn.* **36**, 78–88.

VLEUGEL, D. A. (1962). Über nachtlichen Zug von Drosseln und ihre Orientierung. *Vogelwarte,* **21**, 307–13.

VRIES, H. DE (1948). Die Reizschwelle der Sinnesorgane als physikalisches Problem. *Experientia,* **4**, 205–13.

WALCOTT, C. & MICHENER, M. (1967). Analysis of tracks of single homing pigeons. *Proc. XIV Int. Orn. Cong. Oxford*, 311–29.

WALLIN, L. (1962). Fotografiskt belägg för grågassens halsbrytande akrobatik vid landningskast. *Vår Fågelvärld*, **21**, 133–5.

WALLRAFF, H. G. (1959a). Über den Einfluss der Erfahrung auf das Heimfindevermögen von Brieftauben. *Z. Tierpsychol.* **16**, 424–44.

WALLRAFF, H. G. (1959b). Örtlich und zeitlich bedingte Variabilität des Heimkehrverhaltens von Brieftauben. *Z. Tierpsychol.* **16**, 513–44.

WALLRAFF, H. G. (1960a). Über Zusammenhänge des Heimkehrverhaltens von Brieftauben mit meteorologischen und geophysikalischen Faktoren. *Z. Tierpsychol.* **17**, 82–113.

WALLRAFF, H. G. (1960b). Können Grasmücken mit Hilfe des Sternenhimmels navigieren. *Z. Tierpsychol.* **17**, 165–77.

WALLRAFF, H. G. (1965). Über das Heimfindevermögen von Brieftauben mit durchtrennten Bogengängen. *Z. vergl. Physiol.* **50**, 313–30.

WALLRAFF, H. G. (1966a). Versuche zur Frage der gerichteten Nachtzug-Aktivität von gekägigyen Singvögeln. *Verh. dtsch. Zool. Ges. Jena 1965*, 338–56.

WALLRAFF, H. G. (1966b). Über die Heimfindeleistungen von Brieftauben nach Haltung in verschiedenartig abgeschirmten Volieren. *Z. vergl. Physiol.* **52**, 215–59.

WALLRAFF, H. G. (1966c). Über die Anfangsorientierung von Brieftauben unter geschlossener Wolkendecke. *J. Orn.* **107**, 326–36.

WALLRAFF, H. G. (1967). The present status of our knowledge about pigeon homing. *Proc. XIV Int. Orn. Cong. Oxford*, 331–58.

WALLRAFF, H. G. (1968). Über die Flugrichtungen verfrachteter Brieftauben in Abhängigkeit von Heimatort und von Ort der Freilassung (in prep. ref. in Wallroff, 1967.).

WALLRAFF, H. G. & KIEPENHEUR, J. (1962). Migracion y orientacion en aves: observaciones en Otoño en el sur-oeste de Europa. *Ardeola*, **8**, 19–40.

WALLRAFF, H. G. & KLEBER, N. (1967). Eine Dressurmethode für Wahrnehmungphysiologische Untersuchungen und ihr Anwendung in Orientierungsversuchen mit Vögeln. *Experientia*, **23**. 312–14.

WALLS, G. L. (1963). *The Vertebrate Eye and its Adaptive Radiation*. New York and London.

WALLS, G. L. & JUDD, H. D. (1933). The intra-ocular colour filters of vertebrates. *Brit. Ophthalmol.* **17**, 641–75, 705–25.

WATERMAN, T. H. (1963). The analysis of spatial orientation. *Ergebn. Biol.* **26**, 98–117.

WATSON, J. B. & LASHLEY, K. S. (1915). An historical and experimental study of homing in birds. *Publ. Carneg. Inst. Wash.* **7**, 7–60.

WEHNER, E. & LINDAUER, M. (1966). Zur physiologie des Formenschens bei der Honigbienne. I. Winkelunterscheidung an vertikal orientierter Streifesmustern. *Z. vergl. Physiol.* **52**, 290–324.

WEVER, R. (1967). Über Beeinflussung der circadian Periodik des Menschen durch Schwache electromagnetische Felder. *Z. vergl. Physiol.* **56**, 111–28.

WHITNEY, L. F. (1963). Landmarks and homing. *Amer. Racing Pigeon News*, **79**, 8–10.

WILKINSON, D. H. (1949). Some physical principles of bird orientation. *Proc. Linn. Soc. Lond.* **160**, 94–9.

WILKINSON, D. H. (1950). Flight recorders. A technique for the study of bird navigation. *J. Exp. Biol.* **27**, 192–8.

WILKINSON, D. H. (1952). The random element in bird 'navigation'. *J. Exp. Biol.* **29**, 532–60.

WILLIAMS, C. S. & KALMBACH, E. R. (1943). Migration and fate of transported juvenile waterfowl. *J. Wildlife Manag.* **7**, 163–9.

WILLIAMSON, K. (1955). Migrational drift. *Acta XI Congr. Int. Orn. Basel*, 179–86.

WILTSCHKO, W. & MERKEL, F. W. (1966). Orientierung zugunruhiger Rotkehlchen im statischen Magnetfeld. *Verh. dtsch. Zool. Jena 1965*, 362–7.

WODZICKI, K., PUCHALSKI, W. & LICHE, H. (1938). Untersuchungen über die Orientation and Geschwindigkeit des Fluges bei Vögeln. III. Untersuchungen am Störchen (*Ciconia c. ciconia*). *Acta Orn. Mus. Zool. Pol.* **2**, 239–58.

WODZICKI, K., PUCHALSKI, W. & LICHE, H. (1939). Untersuchungen über die Orientation und Geschwindigkeit des Fluges bei Vögeln. V. Weitere Versuche an Störchen. *J. Orn.* **87**, 99–114.

WODZICKI, K. & WOJTUSIAK, R. J. (1934). Untersuchungen über die Orientation und Geschwindigkeit des Fluges bei Dögeln. I. Experimente an Schwalben (*H. rustica* L.). *Acta Orn. Mus. Zool. Pol.* **1**, 253–74.

WOJTUSIAK, R. J. (1949). Polish investigations on homing birds and their orientation in space. *Proc. Linn. Soc. Lond.* **160**, 99–193.

WOJTUSIAK, R. J. (1960). Orientation in birds. (in Polish, English summary). *Przegladu Zool.* **4**, 254–71.

WOJTUSIAK, R. H. & FERENS, B. (1938). Untersuchungen über die Orientation und Geschwindigkeit des Fluges bei Vögeln. IV. Heimkehrgeschwindigkeit und Orientierungart bei den Rauchschwalben (*H. rustica* L.). *Bull. Acad. Pol. Sci.* **2**, 173–201.

WOJTUSIAK, R. J. & FERENS, B. (1947a). Homing experiments on birds. VII. Further investigations on the velocity of swallows (*Hirundo rustica* L.) and on the role of memory in their orientation in space. *Bull. Int. Acad. Pol.* **2**, 135–64.

WOJTUSIAK, R. J. & FERENS, B. (1947b). Homing experiments on birds. VIII. Observations on the nest, the age and the faculty of orientation in space of chimney swallows (*Hirundo rustica* L.). *Bull. Int. Acad. Pol.* **2**, 165–7.

WOJTUSIAK, R. J., FERENS, B., SCHIFFER, Z., DYLEWSKA, M. & MAJLERT, Z. (1953). Homing experiments on birds. IX. Further investigations on Tree Sparrows, *Passer montanus* (L.). *Acta Orn. Mus. Zool. Pol.* **4**, 311–34.

WOJTUSIAK, R. J., WODZICKI, K. & FERENS, B. (1937). Untersuchungen über die Orientation und Geschwindigkeit des Fluges bei Vögeln. II. Weitere Versuche an Schwalben: Beeinflussung durch Nachtzeit und Gebirge. *Acta Orn. Mus. Zool. Pol.* **2**, 39–61.

WOJTUSIAK, R. J., WOJTUSIAK, H. & FERENS, B. (1947). Homing experiments on birds. VI. Investigations on the Tree and House Sparrows (*Passer arboreus* Bewick and *P. domesticus* L.). *Bull. Int. Acad. Pol.* **2**, 99–106.

Wolfson, A. (1966). Environmental and neuroendocrine regulation of annual gonadal cycles and migratory behaviour in birds. *Recent Progress in Hormone Research*, **22**, 177–244.

Yeagley, H. L. (1947). A preliminary study of a physical basis of bird navigation. *J. Appl. Phys.* **18**, 1035–63.

Yeagley, H. L. (1951). A preliminary study of a physical basis of bird navigation. II. *J. Appl. Phys.* **22**, 746–60.

Yule, G. U. (1926). Why do we sometimes get nonsense-correlations between time series? *J. Roy. Statistical Soc.* **89**, 1–58.

Zeigler, H. P. & Schmerler, S. (1965). Visual discrimination of orientation by pigeons. *Anim. Behav.* **13**, 475–7.

# INDEX

accelerometers, 97
activity, period, 49, 156, 157
  threshold, 42, 156
adult, 1, 9, 14, 17, 58, 65
air sacs, 95
Albatross, Laysan, 60, 62, 63, 69
almanac, nautical, 38, 53, 120
amphibian orientation, 52
anaesthesia, 98
ancestral home, 18, 59, 163
'angels', 5
arteries, 104
arthropod, orientation, 21, 26, 27, 29,
  33, 35, 36, 38, 39, 52, 85, 142, 145,
  146, 149
  clocks, 33, 159, 160
artificial horizon, 152
  satellite, 75
  sun, 28–30, 141, 142, 148
atmospheric conditions, 79, 95, 110
auditory clues, 5, 24, 27, 138, 155, 156
automatic registration, 22, 24, 92
aviary experiments, 21, 59, 87–9, 135–7

barometric pressure, 27, 111
bats, orientation of, 156
bearing-and-distance, 3, 10, 17, 162
bi-component hypothesis, 39, 155
Blackbird, 147
  Red-winged, 60, 62
  Yellow-headed, 59
Blackcap, 23, 41, 42, 47, 48, 93, 117
blind-folded birds, 154
Bluethroat, 60
Brambling, 23
breeding, area, 1–3, 9, 17, 48
  cycle, 7, 56, 59, 65
Brownian energy, 104
Bunting, Indigo, 42, 47, 48
  Ortolan, 42
  Reed, 23, 147

cages, orientation, see Kramer
carry-over orientation, 24
castration, 56
ceilometer, 4

centrifugal force, 102
cerebellum, 99
Chaffinch, 4, 8, 13, 23, 27, 93, 147
chickens, 110, 154, 155
chilling of eggs, 59
chronometer, 119, 122, 138, 156, 160,
  162
circular choice apparatus, 23, 28–31,
  33, 42
circumpolar constellations, 44, 48–51
Cisticola, Chubb's, 156
clock, biological, 30–32, 156–60, 162
  annual, 160, 161
  lunar, 52
  shifting, 28–33, 38, 44, 45, 52, 53, 84,
    85, 138–41, 159–60
closed room, orientation in, 24, 25, 91,
  92
cloud, see overcast
comparison of stimuli, 101, 102, 120,
  122–6, 130, 132
compensation orientation, 42, 47
conflict orientation, 38, 42, 48
conjugate point, 107–9
Coot, 57
cosmic rays, 158
couple, fluid, 103
courtship signals, 155
Cowbird, 60
Crossbill, 23
  Parrot, 23
cross-over point, 123, 127
Crow, Prairie, 10
  Hooded, 13, 15–17
Cuckoo, 10
  Black-billed, 42
  Bronze, 2
  Yellow-billed, 42
culmination, 119, 122, 123, 127, 130,
  132, 134, 161
Curlew, Bristle-thighed, 1
curvature of earth, 73, 94

dark adaptation, 150
day length, 31, 32, 38, 39, 43, 136, 137
decompression, 99

delay experiments, 10, 15
detour experiments, 96
dichotomy in homing, 88, 89
direction maintenance, 8, 9, 23, 24, 27,
    41, 43, 115
directional bias, innate, 12, 15, 18, 21,
    92
    learned, 10, 18, 27, 30, 35, 92
direct sensory contact, 94–100
disorientation, 7, 8, 24, 25, 28, 38, 42,
    43, 48, 88, 90, 91, 114, 138
    zone of, 82, 83, 86, 90, 94, 101,
    111–13, 140, 153
displacement of migrants, 13–17, 54,
    162, 163
distance, of migration, 15
    of transportation, 82–4, 94, 102, 114,
    119, 140, 148, 156
Double Altitude solution, 122
Dove, Mourning, 3
doves, migratory, 99
drum, rotating, 98
Duck, Wood, 13, 59
ducks, 3, 12, 42, 68, 154, 162
duetting, 155, 156
Dunnock, 42, 57

Eagle, Bald, 72
    Golden, 146
echolocation, 156
eclipse of sun, 114
'electric, braes', 152
electrodes, implanted, 99, 143–4
electromagnetic field, 25, 105, 106, 110
electrostatic field, 26, 105
equator, 32, 36, 37, 49, 129, 145
equatorial, trans-, experiments, 35, 36,
    129
equinox, 33, 36, 122, 129, 133, 134
evolution, 35, 36, 50, 56, 67, 152, 158,
    163
experience, role of, 9, 18, 19, 30, 36, 56,
    65, 86, 88, 89, 116
extrapolation, 30, 112, 123–7, 130, 144,
    145, 155
extrasensory perception, 95, 96
'eye-sign', 144
eye structure, 143, 144, 146, 147, 150,
    152

feed-back systems, 98, 154
Fieldfare, 147
fish, orientation of, 26, 33, 35, 36, 38, 39
flicker-fusion, 143

flight, recorder, 68, 69
    speed, 65, 97, 103, 105, 106, 128, 155
    vector, 108, 140, 141, 162
flocks, behaviour of, 20, 63, 72, 114,
    118
Flycatcher, Collared, 59
    Pied, 3, 59, 93
Flying Squirrel, 156–8
'flyways', 58
fog, 4, 7, 8
food, training to, 56, 57, 87, 158–9
foster home, 59
Frigate Bird, 57

Gannet, 60, 71, 72, 144
gastropods, orientation of, 26
geese, 152, 153, 162
Goldeneye, 155
Goldfinch, 28
Gonolek, Black-headed, 155
Goose, Barnacle, 3
    Canada, 12, 20, 27, 43, 69
Goshawk, 57
gravity, 97, 102, 103, 136, 152–4
Greenfinch, 23
grid, navigational, 101–31
Grosbeak, Rosy, 42
    Scarlet, 23, 42, 93
Gull, Black-headed, 57, 61, 91, 92
    Common, 59
    Herring, 60, 62, 64, 66, 69, 107
    Lesser Black-backed, 60, 61, 69, 89,
    107, 112, 113
gyroscope, 97

hand-rearing, 41, 92, 117
Hawfinch, 23
Hawk, Sparrow, 13, 17
head stability, 100, 103, 153
heart beats, 47
homing, 17, 54–93, 99, 104, 108, 116,
    121, 129, 132–41
    night, 118, 119
    speed, 63–67, 84, 116, 121, 129, 139,
    141
homeward component, 83, 84, 136, 140
    orientation, 67–72, 76–93, 112, 115,
    116, 121, 129, 135, 151
horizon, 89, 124, 135–7, 152
hormones, 31, 56
hourglass timing, 52
human capacities, 123, 142, 143, 147,
    149, 150, 152, 153, 156
humming birds, 155

immediate orientation, 78–82
imprinting, of locality, 59
individual, capacities, 43, 114, 151
    variation, 55, 65, 88, 159
inertial navigation, 96–100
infra-red, photography, 23
    sensitivity, 95

juveniles, 1, 8, 10–17, 42, 56–9, 85–87,
    162, 163

Kestrel, 153
kinesis, 117
Kramer-cage, 22–31, 35, 41, 42, 47, 51,
    90–3, 117, 118

labyrinth, destruction of, 98, 99, 155
lamps, bird-borne, 43, 118
landmarks, 9, 18, 20, 23, 27, 55, 57, 65,
    73, 82, 84–92, 110, 128, 135, 140,
    141
Lark, Western Meadow, 29
latitude, 35–9, 48, 97, 112, 117, 119,
    125–8, 132
leading lines, 3, 72
light, constant, 31, 158, 159
light/dark cycle, 31, 138, 158
lighthouses, 4
Linnet, 23
loft, mobile, 107
    situation, 73, 85, 86, 134
longitude, 32, 47, 48, 112, 117, 119,
    122, 125, 127, 138, 140, 141, 156,
    159

magnets, bird-borne, 25, 105–7
magnetic field, 25–27, 104–9, 158
Mallard, 12, 13, 18–21, 25, 27, 29, 31,
    33, 43, 45, 48, 52, 53, 58, 61, 72, 79,
    81, 84, 119, 150
map-and-compass, 85, 102, 110, 111,
    117
Martin, House, 60
    Purple, 60, 61, 62, 118
mass, measurement of, 102, 103
memory, 87, 130, 132, 137, 160–3
meredian, 43, 44, 119, 122
migration, forms of, 1–7, 24, 41
    restlessness, 17, 22, 23, 25, 117
    stimulus, 22, 23, 117, 118
minimum separabile, 147–149
mirror experiments, 28, 29, 38, 85
moon orientation, 8, 41, 49, 51–3, 149
    -watching, 5

moult, 56, 68
movement detection, 142–4, 152

nasal cavities, 95
non-migratory stock, 12, 13, 65, 85, 94,
    99
'nonsense orientation', 18–21, 27, 29,
    31, 32, 43, 52, 78–82, 84, 85, 88,
    89, 127, 140
noon, 38, 112, 122, 126, 130, 131
nutation, 49

obliques, discrimination of, 149
Oil Bird, 156
oil droplets, retinal, 146
one-direction navigation, 22–25
operant conditioning, 25, 142, 148, 159,
    161
optomotor reaction, 142
orbit, of earth, 44, 50
Ortstreue, 2
oscillatory systems, 33, 34, 39, 97, 158
Ostrich, 147
Ousel, Ring, 15
overcast, effect of, 4, 7, 8, 27, 29, 42, 43,
    46, 74, 84–6, 91, 92, 112–17,
    138–41, 150, 159
owls, 95

palisade experiments, 88, 89, 135–7
parallax, 47, 124, 135, 136, 145
paratympanic organ, 99
pecten, 105, 143
Penguin, Adelie, 21, 27, 32, 60, 68
Petrel, Fulmar, 150
    Leach's, 60–63, 66, 98, 106
    Storm, 60
phototaxis, 51, 85, 91
Pigeon, homing, 17, 18, 25, 27, 29–33,
    35–8, 54–7, 59, 64–5, 68–72, 77–90,
    94, 95, 98, 99, 102, 106–10, 114–16,
    118, 127, 128, 133–42, 147–50, 161
    Rock, 65
pigeon-racing, 17, 54, 56, 71, 108
pinhole camera, 121
Pintail, 20, 21, 27, 43, 58, 61, 119
Pipit, Meadow, 23
    Tree, 42
planarians, orientation of, 26
planetarium, 47–9, 117, 118
planets, 49, 50
Plover, Golden, 92, 115
plumage marking, 10, 13
polarised light, 29

Pole, geographical, 21, 37, 107, 125
  Star, 44, 45, 47, 49, 119
position, ambiguity of, 107, 129, 130
  triangle of, 120
postural sensors, 104, 152–5
potential difference, 105, 106
precession, 49, 50
pressure-pattern flying, 26
proprioception, 96
protozoa, orientation of, 26

Quinine hydrochlorate, 99

radar, 5–8, 72, 110
radio-active watcher, 61
radio-transmissions, 109, 110
radio-transmitters, bird-borne, 43, 72–5, 114, 141
random scatter, 42, 86, 107, 108, 114, 134, 140
release site, 76, 108, 110, 135
re-orientation, 8, 17, 48
residual orientation, 24
retina, 143–50, 152, 155
retracement theories, 96
retrapping, 59, 70, 71
rhythms, biological, 31, 155, 160, see also clocks
ringing recoveries, 2, 10–13, 58, 69, 70, 108, 109, 135–7
rings, leg, 2, 10, 60
Redstart, 93
Robin, 24, 25, 42, 147
  American, 150
rotating drum, 98
rotation of earth, 26, 43, 44, 97, 103, 104, 119, 129

sea, release at, 64, 89, 124, 125
search, random, 67–8, 72, 94, 98, 107
  systematic, 66, 72
seasonal changes, 36–40, 43–50, 65, 110, 111, 126, 127, 132–4, 161, 162
selection, 54, 65
semi-circular canals, 96, 99, 100, 104, 106, 152
sensory summation, 146, 150
sextant, 123, 135
Shearwater, Great, 1, 3, 161, 162
  Manx, 59–65, 69, 81, 89, 90, 112, 113, 116, 118, 138, 150
  Short-tailed, 160–2
  Sooty, 3
Shelduck, 59

Shrike, Lesser Gray, 41
  Red-backed, 23, 41, 42, 60, 61
sightings, en route, 69
Siskin, 23, 158
Skua, Great, 2
sky-glow, 41, 95
sky-watching, 4
smell, sense of, 95, 150
soaring, 63, 72
solstice, 37, 133, 134
Sparrow, Eastern Tree, 57
  Fox, 2
  Golden-crowned, 58, 62
  House, 57, 62, 64, 131, 150
  Tree, 57, 59, 62, 64
  White-crowned, 41, 58, 62, 64
stabilised platform, 97
standard direction, 3, 9, 17
star, altitude, 43, 119, 120
  azimuth, 43, 45, 47, 50, 120
  bi-coordinate navigation, 112, 117–20
  -compass, 43, 48, 105, 119
  pattern, 8, 41, 42, 43, 47, 112
Starling, 13–16, 23, 27–31, 35, 36, 60, 61, 66, 69, 89, 90, 93, 95, 98, 147, 150, 158, 159
statistical analysis, 24, 25, 49, 59, 61, 76, 91, 134
steeple-chasing, 73, 86
stock, variations in, 55, 64, 65, 81
Stork, White, 10–12, 60, 61, 69, 107
Sumner circles, 118–20, 127, 128
sun, altitude, 30–9, 112, 122–30, 133, 148, 149
  instantaneous, 127–9
  rate of change of, 36–9, 125
arc, 37, 122, 124, 125, 127, 132, 133, 148, 161
  inclination of, 122, 125, 133, 148, 161
azimuth, 36–40, 112, 123, 124, 130, 145
  rate of change of, 36–9, 125
bi-coordinate navigation, 117, 122–42
-compass, fixed angle, 9, 29, 30, 163
  time-compensated, 26–36, 41, 80, 85, 86, 92, 105, 110, 115, 116, 132, 138–41
movement, direction of, 129, 130, 145, 148
nocturnal position, 33, 34, 39
sunrise, 8, 112, 130, 136, 159, 160
sunset, 41, 112, 130, 136

sunspot activity, 108
Swallow, 57, 60, 61, 66, 110, 118
 Bank, 3, 60, 81, 90
 Cliff, 60
Swift, 60, 147
 Alpine, 60, 61
Swiftlet, 60, 147

Teal, Blue-winged, 10, 15, 20, 27, 43
 Green-winged, 13, 14, 20, 58
television masts, 4
Tern, Arctic, 2, 60, 61
 Common, 18, 27, 60, 66
 Noddy, 60, 62, 95
 Sooty, 60, 62, 69
territorial behaviour, 94
threshold-tracking, 148
Thrush, Song, 147
 Grey-cheeked, 74, 75
 Swainson's, 74
Thrush-Nightingale, 42
thunderstorms, 108
time, 29, 30, 36–8, 40, 44, 119, 124
 local, 39, 122, 127, 129, 137
time-interval measurements, 145, 155,
 156, 159, 161
time-shift, 32, 47, 48, 160
topographical features, 8, 9, 19, 20, 56,
 72, 73, 108
tracking, from ground, 68, 72, 73
 by aircraft, 18, 43, 71, 73–5, 107,
 108, 114, 141
training, directional, 17, 23, 27, 30, 35,
 42, 47, 55, 73, 78, 82, 86, 114, 115,
 135, 141, 158
 non-directional, 55, 82, 135
transplantation experiments, 10–13
turbulence, 7, 8, 9, 153, 154

turntable experiments, 46, 47, 98, 99
twilight, 42

ultra-violet light, 27
unknown senses, 94, 95, 96, 110
untrained Pigeons, 78, 82, 85, 87, 89,
 135–7

vanishing time, 76–80, 114
vertical, determination of, 149, 152
vision, long distance, 94, 95
 night, 149–51
visual, acuity, 146–8, 151
 observation 4, 7
 reference points, 8, 9, 27, 115, 124,
 154

Wagtail, White, 23
waiting experiment, 133–6
Warbler, Barred, 23, 27, 41, 42, 93
 Garden, 23, 41, 42, 47, 48, 117
 Grasshopper, 42
 Wood, 23, 41
wave patterns, 7, 125
Whinchat, 93
Whitethroat, 24, 41, 117
 Lesser, 41, 47, 48, 117
wind, effects of, 3, 6, 7–9, 19, 26, 90–2,
 108, 109, 112, 128, 154
winter, effect on homing, 111
wintering area, 1, 2, 4, 9, 12, 17, 54, 57,
 58
Wryneck, 60, 71

year length, 50, 160–61
Yellowhammer, 147

*Zeitgeber*, 31, 138, 160
zenith, 38, 119, 121, 145